Principles of Biology │ Cells

Biology Lab Manual

First Edition

Jennifer Schramm

Chemeketa Press │ Salem, Oregon

Principles of Biology | Cells: Biology Lab Manual
© 2025 by Jennifer Schramm

ISBN-13: 978-1-955499-45-3

Chemeketa Press
Chemeketa Community College
4000 Lancaster Dr NE
Salem, Oregon 97305
collegepress@chemeketa.edu
chemeketapress.org

Cover design by Ronald Cox
Interior design by Ronald Cox and Abbey Gaterud

A full list of credits appears on page 153 and constitutes an extension of this copyright page.

References to website URLs were accurate at the time of writing. Neither the author nor Chemeketa Press is responsible for URLs that have changed or expired since the manuscript was prepared.

Printed in the United States of America.

Land Acknowledgment
Chemeketa Press is located on the land of the Kalapuya, who today are represented by the Confederated Tribes of the Grand Ronde and the Confederated Tribes of the Siletz Indians, whose relationship with this land continues to this day. We offer gratitude for the land itself, for those who have stewarded it for generations, and for the opportunity to study, learn, work, and be in community on this land. We acknowledge that our College's history, like many others, is fundamentally tied to the first colonial developments in the Willamette Valley in Oregon. Finally, we respectfully acknowledge and honor past, present, and future Indigenous students of Chemeketa Community College.

Contents

Lab and Safety Regulations

Personal Behaviors

1. Eating and drinking are prohibited in the labs.
2. Wash your hands before leaving the lab.
3. Wash the lab counter & related work areas with disinfectant before and after the lab.
4. Stow your personal items in the cubbies or under your station.
5. Keep the lab counters and work areas uncluttered.
6. Clean-up is your responsibility:
    a. Return cleaned items back to lab kits.
    b. Wash slides and return to the original location (unless told otherwise). Cover slips may be thrown away.
    c. Wash glassware and return to original location (unless told otherwise).
    d. Ensure any lab waste is disposed of in the approved container(s).
    e. Ensure sinks and counters are as clean as when you arrived.
7. Clothing must be appropriate for the lab to be performed.
8. Read the lab in advance so you're aware of proper safety protocols.
9. If you have questions about safety, ask before you do.

Safety Protocols

1. Know where the lab safety equipment is located—fire extinguisher, first aid, eye wash, etc.
2. Wear personal safety equipment (goggles, gloves, aprons) as indicated in the lab instructions or by your instructor.
3. Handle all chemicals and biologicals, including stains, below eye level.
4. Do not assume something can just go down the sink or in the trash. Dispose of wastes in appropriate waste containers.
5. If there is a chemical spill:
    a. Get your instructor.
    b. If the spill is on the floor or counter, keep away from it.
    c. If the spill is on you or your clothes, rinse the area immediately with running water.
    d. If you splash chemicals in your eyes, go immediately to the eye wash station, turn on cold water, remove the red caps and lean down so that the water bubbles into your eyes.
6. In case of injury:
    a. Get your instructor.
    b. If the instructor is not available, call campus safety (x5023) on the lab's phone.
7. If the injury appears severe, also call 911
8. In case of fire:
    a. If your clothes are on fire, yell "FIRE" and roll on the floor or use a coat or fire blanket to smother it.
    b. If chemicals or lab equipment is on fire, call out "FIRE" and evacuate the room.
9. If the fire alarm goes off, take your essential items and leave the room. Head to the evacuation area as told by your instructor. Stay with your class and wait for instructions. Do not re-enter the building until an all-clear is given.

1 | Diffusion & Osmosis Pre-Lab

Instructions

- ❑ Read the lab manual and then follow the instructions to complete the assignment.
- ❑ You may need your textbook and other resources to complete this assignment.
- ❑ Pre-labs must be completed before the start of the lab.

1. Differentiate between the terms diffusion and osmosis.

2. During which of these processes (diffusion or osmosis) are the terms isotonic, hypotonic, and hypertonic applied?

Figure 1.1 shows three different examples of two solutions being separated by a selectively permeable membrane (dotted line in the middle). The membrane is not permeable to sucrose or starch.

Figure 1.1. Isotonic, hypertonic, or hypotonic solutions.

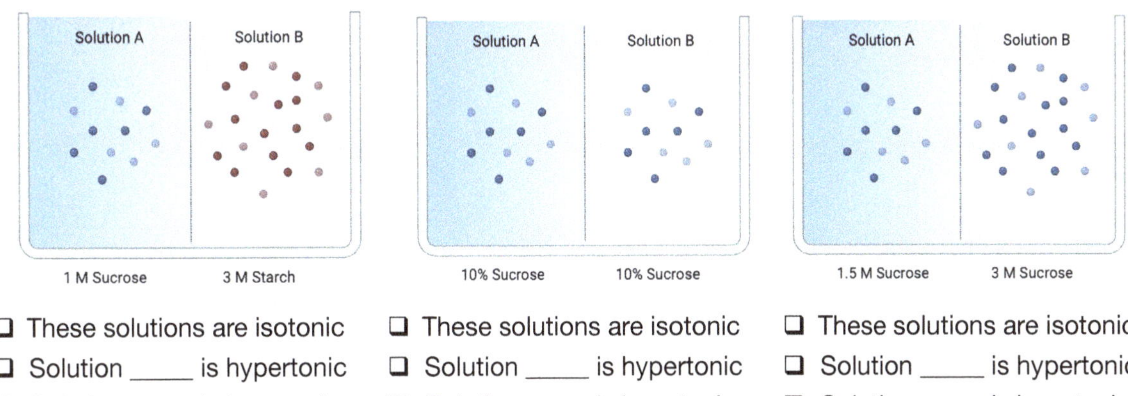

- ❑ These solutions are isotonic
- ❑ Solution _____ is hypertonic
- ❑ Solution _____ is hypotonic

- ❑ These solutions are isotonic
- ❑ Solution _____ is hypertonic
- ❑ Solution _____ is hypotonic

- ❑ These solutions are isotonic
- ❑ Solution _____ is hypertonic
- ❑ Solution _____ is hypotonic

Instructions

1. Check the correct box below each set of solutions presented in Figure 1.1 and, if needed, fill in the blanks to indicate which solution is hypertonic or hypotonic.
2. Draw an arrow on each diagram indicating the direction of water movement in that set of solutions.
3. Review the experiments we will perform this week and complete Table 1.1 by providing a valid hypothesis for each experiment.

3. Do a little research (don't spend too much time on this). What is the average concentration of solutes in a cell? *Keywords: cytoplasm/ic, cytosol/ic, concentration, solute*

Remember that a hypothesis:
- ❑ *Answers the question being asked*
- ❑ *Does not suggest the experiment that will be performed*
- ❑ *Does not predict the results of an experiment*

Table 1.1. Creating a Valid Hypothesis for Each Experiment

Exercise	Question	Hypothesis
2	What factors change the rate of diffusion of food coloring?	
3	How can osmosis be used to determine the concentration of solutes in the cytoplasm of a cell?	

1 Diffusion and Osmosis

By the end of this lab, you should be able to:
- ❑ Describe how osmosis impacts the shape of a cell.
- ❑ Design an experiment to determine the concentration of the cytosol of the cells of a particular tissue.
- ❑ Explain why cell membranes are considered selectively permeable.
- ❑ Explain why diffusion and osmosis occur (at the molecular level).
- ❑ Use the terminology associated with osmosis (hypotonic, hypertonic, isotonic).

> Many activities in this lab require 30 minutes to 2 hours to complete. Set up these activities in the following order and work on the remaining activities while they are running.
> - ❑ Activity 2 (Potato Osmosis)
> - ❑ Activity 1, Procedure A (Diffusion and Molecular Mass)
> - ❑ Activity 1, Procedure B (Diffusion and Temperature)
> - ❑ Activity 3 (Animal Blood Osmosis)-set up is quick, but interpretation is challenging

Activity 1. What Factors Affect the Rate of Diffusion?

Many factors affect the rate of diffusion, and we will explore the impact of these factors on cells throughout the term. In this exercise, we are going to focus on the effect of molecule size, temperature, viscosity, and surface area on diffusion. The rate of diffusion is also affected by differences in concentration gradient and electromagnetic attraction.

To model the rate of diffusion in different-sized cells, perform a set of experiments using agar, a gelatinous extract from seaweed. The gelatinous features of agar make it similar to the cytosol of the cell. While the cytosol is water-based, the macromolecules within make it more gelatinous. We will use food dyes as our solutes for these experiments. Food coloring is made of a combination of a few different molecules. You can determine what dyes are in your food coloring by examining the ingredients on the bottle.

Procedure A. Solute Mass (2 hours)

The mass of a solute is determined by adding up the masses of the elements that make up that molecule. Since different food dyes have varying masses, we can use them to study the effect of mass on the diffusion process.

Q1. **Hypothesis:** How do you think solute size affects the rate of diffusion?

Q2. **Determining Mass:** Assess the ingredients list of two different colored food dyes to determine which dyes are present. Use the chemical formula from the table provided in class to calculate the molecular weight of the dyes. Show your work below.

Q3. **Prediction:** Which dye do you think will travel fastest based on your hypothesis?

Instructions

1. Record the names and masses of the dyes you are using in the appropriate cells of Table 1.2.
2. Use a marker to draw a line on the bottom of a petri dish containing agar, dividing it in half. Label the halves A and B.
3. In the center of each half, press the straw into the agar, twist, and pull to remove a "plug" of agar. This will create an empty well in the center of each side of the dish.
4. Use a pipette to add a drop of food coloring to the well in half A.
5. Repeat the process for half B and the other food coloring.
6. Return the lid to the dish.
7. Use the procedure below to measure the distance each dye diffused in millimeters (mm) every 30 minutes for 2 hours. If your food coloring includes more than one color, follow the one that moves the fastest.
    a. Turn the dish upside down for the clearest image.
    b. Put a ruler on top of the dish.
    c. Measure the distance from the edge of the well to the "dye front."
8. Record your results in Table 1.2.
9. Calculate the rate of diffusion at each time interval and average the rates together for each dye to determine the average rate of diffusion per hour. Record the averages in Table 1.2.

Table 1.2. The Diffusion Distance of Differently Massed Dyes in Agar

Time (min)	Distance Traveled (mm)	
	Dye 1: MW:	Dye 2: MW:
30		
60		
90		
120		
Average Rate (mm/hour)		

Q4. How does solute mass affect the rate of diffusion?

Procedure B. Temperature (1 hour)

Temperature is a measurement of the amount of kinetic energy stored in a molecule. As that energy increases, the temperature increases.

Q5. **Hypothesis:** How do you think solute size affects the rate of diffusion?

Q6. **Prediction:** Which temperature do you think will exhibit the fastest rate of diffusion based on your hypothesis?

Instructions

1. Set a heating block to 85°C. Once warm, move a warm agar plate from the class stock to your heating block.
2. Obtain a cold agar petri dish from the refrigerator.
3. Use a straw to create one well in the center of each petri dish.
4. Put a drop of food coloring (one of the ones you used in part A) into the well of each dish.
5. Return the dishes to the refrigerator or heating block.

6. Use the procedure below to measure the distance each dye diffused in millimeters (mm) every 15 min for the next hour. If your food coloring includes more than one color, follow the one that moves the fastest.
    a. Turn the dish upside down for the clearest image.
    b. Put a ruler on top of the dish.
    c. Measure the distance from the edge of the well to the "dye front" and record your measurements in the table you created in your lab notebook.
7. Record your results in Table 1.3. Complete your table by adding the room temperature results for that dye from Table 1.2.
8. Calculate the rate of diffusion at each time interval and average the rates together for each dye to determine the average rate of diffusion per hour. Record the averages in Table 1.3.

Table 1.3. Distance of Dye Diffused in Agar at Different Temperatures

Time (min)	Distance Traveled (mm)		
	Cold	Room Temperature	Hot
15			
30			
45			
60			
Average Rate (mm/hour)			

Q7. How does temperature affect the rate of diffusion?

Procedure C. Viscosity

Viscosity is a measurement of a liquid's resistance to flow; in other words, how thick the liquid is. Water is considered to have low viscosity, and honey has high viscosity.

Q8. **Hypothesis:** How do you think viscosity affects the rate of diffusion?

Instructions

1. Fill the lid of a petri dish with a layer of tap water and allow it to settle on your workbench for a few minutes.

2. Add a drop of food coloring carefully to the top of the water, trying not to disturb the surface.
3. Use your observations to draw a conclusion.

Q9. How does viscosity affect the rate of diffusion?

Procedure D. Surface Area

The size of a cell is limited by the amount of key materials (solutes) that the cell can absorb. That value is determined by the surface area (region exposed to the environment) of the cell. In this experiment, we use the pH-sensitive dye, bromothymol blue, to study the impact of surface area.

Instructions

1. To see what happens when this dye is exposed to vinegar (acetic acid), put a drop of dye and a drop of vinegar on a white surface.
2. Record your observations.

Q10. **Hypothesis:** How does surface area affect the rate of diffusion?

3. Obtain a cube of blue agar. Use a razor blade to cut three smaller blocks of obviously different sizes from that cube.
4. Use a metric ruler to measure the size of one side of the cube in millimeters (mm) and record your measurement in Table 1.4.
5. Complete Table 1.4 as follows:
    a. Perform calculations as indicated in the table.
    b. Also record the volumes in Column A of Table 1.5.
    c. Indicate the surface area: volume ratio for each block (reduce as you would a fraction).
    d. Predict which cube will exhibit the fastest rate of diffusion.
6. Fill a beaker with around 3 cm of distilled white vinegar (enough to cover your largest cube).
7. Place all three cubes into the container for 5 minutes.
8. Remove the cubes from the vinegar and place them on a white surface.
9. Use a ruler to measure the length of one side of the part of the cube that remains blue (where the vinegar has not penetrated).
10. Record your measurements in Table 1.5 and complete calculations as described in the table.

Table 1.4. Surface Area Volume Ratio of Agar Blocks

Block	Length of one side (mm)	Surface area (6 X L X W)	Volume (L X W X H)	Surface Area to Volume Ratio (SA : V)	Prediction (rank speed with #1 being fastest)
Large					
Medium					
Small					

Table 1.5. Determining Vinegar Penetration in Agar Blocks

Column	A	B	C	D	E	F
Block	Volume of original block (Table 1)	Length one side of blue region after diffusion (mm)	Volume Blue region after diffusion (L X W X H)	Volume of Area that underwent diffusion Column A – Column C	Fraction of cube that underwent diffusion Column D/ Column A	% Cube Penetrated with Vinegar Column E X 100
Large						
Medium						
Small						

Optional Activity. How Does the Surface Area of an Object Affect the Rate of Diffusion?

Instructions

1. Work with your group and left over pieces of blue agar to create a shape that is about the same size as your medium block, but that has either more or less surface area than the cube.
2. Drop your shape into the vinegar at the same time as everyone else in your group.
3. After 5 minutes, remove your shape and compare it with other group members.

Activity 2. Using Osmosis to Measure Cytosolic Solute Concentration (2 hours)

Water moves in and out of cells based on the amount of solute in the environment versus the amount of solute in the cytosol. When we study osmosis, we use the concentration of two solutions separated by a membrane (like the plasma membrane) to determine which direction water will move. If the solute concentration of those two solutions is the same, we say that the solutions are **isotonic**.

In this case, although water moves across the membrane by natural processes, the net movement of water is zero. If, however, the concentration of the solutions on each side of the membrane differs, we call the solution with higher solute concentration **hypertonic** with regards to the solution with lower concentration (**hypotonic**). In this case, the net movement of water would be from the hypotonic solution to the hypertonic solution.

Water is not compressible and tends to be pretty heavy, so when it moves in or out of the cell, the size and weight of the cell can potentially change as a result of osmosis. We can utilize this potential change in weight due to osmosis as a means to indirectly measure the concentration of the cytosol within a cell.

Q11. How do you use osmosis to determine the concentration of the cytoplasm of a cell?

Q12. **Hypothesis:**

Q13. **Prediction:**

Instructions

1. Use a wax pencil to label each of the 250-ml beakers in your kit with the concentration of the sucrose solution it will contain.
2. Add 100 ml of the indicated solution (water or sucrose) to each beaker.
3. Separate the cork borer from the punch. Push the cork borer through the longest axis of your potato. Twist the cork borer back and forth to loosen it, and then pull it out of the potato.
4. Use the punch to push the cylinder of potato out of the cork borer. Place the core into a petri dish and cover.
5. Repeat until you have seven cylinders of potato.
6. Line the cylinders up and use a fresh razor blade to cut them to the same length. None of the potato skin should remain on the cylinders.

7. Place a plastic weigh boat on the digital scale and hit the tare button. Make sure that the scale is set to measure grams, not ounces.

8. Form an assembly line with your teammate to add the potato to each solution as follows:

    a. Team Member 1: Remove a cylinder from the petri dish and blot it dry with a paper towel. Place the cylinder in a weight boat on the scale.

    b. Team Member 2: Record the initial weight in the appropriate cells of Table 1.6.

    c. Team Member 3: Use the razor blade to cut the cylinder in half lengthwise (like a hot dog bun). Drop the two pieces of the potato into the beaker containing the first solution.

    d. Team Member 4: Record the time: _________.

9. Repeat this procedure until a cylinder of potato has been added to each beaker.

10. Swirl each beaker every 15 minutes.

11. After 1.5 to 2 hours, use the assembly line approach to determine the final weight of each potato:

    a. Team Member 1: Remove the potato pieces from the beaker and blot them dry with a paper towel. Place the potato in a weight boat on the scale.

    b. Team Member 2: Record the final weight in the appropriate cells of Table 1.6.

    c. Team Member 3: Dispose of the potato and re-tare the scale.

12. Record the final weight in the appropriate cells of Table 1.6 and calculate weight change and % weight change as indicated.

Table 1.6. Results of Concentration of Potato Cell Cytoplasm Experiment

	Pure Water	0.1 M Sucrose	0.2 M Sucrose	0.3 M Sucrose	0.4 M Sucrose	0.5 M Sucrose	0.6 M Sucrose
Starting weight (g)							
Final Weight (g)							
Weight change (g)							
% change in weight							

Q14. Which solution caused the least percent weight change in the potato sample?

Q15. How is your answer to the previous question related to the concentration of the cytosol of potato cells?

Q16. Does your data support your hypothesis? Why or why not?

Activity 3. Osmosis in Cells

One of the unique features of animal cells is their complete lack of an external structural feature like the cell wall. Lacking a cell wall gives animal cells a noticeable advantage in terms of flexibility, but also makes them more susceptible to shape changes as a result of osmosis.

In this activity, we will examine how solutions of different concentrations affect the shape of red blood cells. For this experiment, red blood cells were placed in different solutions (A, B, and C). It is your job to determine whether each solution is isotonic, hypertonic, or hypotonic to the cytosol of the red blood cell.

Procedure A. Examining Citrated Ox Blood

Examine the demonstration set up by your instructor. Five drops of citrated ox blood were added to tubes containing solution **A**, **B**, or **C**.

Instructions

1. Hold each test tube in front of a page from your lab.
2. Can you read the print? Record your observations in Table 1.7.
3. Set up four compound microscopes side-by-side on your lab bench.
4. Create 4 wet mounts as follows:
     a. Label the slides A, B, C, and "blood alone" with a wax pencil.
     b. Place a drop of pure blood on the "blood alone" slide, cover with a coverslip.
     c. Place a drop of solution A on slide A and cover with a coverslip. Add a drop of blood to the side of the coverslip. The cells will be drawn under the coverslip by capillary action.
     d. Repeat the previous step to prepare slides B and C.
5. Examine the four slides with a compound microscope, moving between microscopes, and noting differences between them. Add your observations about cell appearance to Table 1.7.
6. Complete Table 1.7 by indicating whether the solutions were isotonic, hypertonic, or hypotonic to the cytoplasm of the blood cell.

Table 1.7. Observations of Ox Red Blood Cells in Solutions A–C

	Appearance of sample containing cells and solution	Can you read the print through a sample? containing cells and solution?	Appearance of cells under microscope (swollen, normal, shrunken)	Is the solution isotonic, hypertonic or hypotonic to cytosol?
Cells Alone	N/A	N/A		N/A
Solution A				
Solution B				
Solution C				

Q17. Explain why we put our blood solutions in front of a page of text? What determines whether you can read the text through the solution?

Q18. If our cells started out in a shrunken state, would we still be able to do this experiment? What would happen?

1 | Applying What You've Learned

A selection of these questions will appear on your lab quiz. Lab quizzes are open book. Type up answers to these questions in advance so that you can copy and paste the answers into the quiz.

1. More is always better when it comes to fertilizing your houseplants, right? Horticulturists know that this is not the case.
2. Use the concepts of osmosis to explain why the application of highly concentrated fertilizer can kill a plant.
3. Carbonated beverages like soda get their bubbles from carbon dioxide, a gas. When you leave a bottle of soda sitting on the counter, the number of bubbles in the soda decreases (the soda goes flat).
4. Is this an example of osmosis or diffusion? Explain your answer.
5. When you turn a cucumber into a pickle, you put it in a salty solution. The end result is a slightly shriveled cucumber that tastes a bit salty.
6. Is pickling caused by osmosis or diffusion (or both)?
7. Is pickling cause by osmosis or diffusion (or both?) Explain your answer.
8. Animal cells lack a cell wall and exhibit much more dramatic shape change in response to hypertonic and hypotonic solutions. The picture below shows images of red blood cells along a concentration gradient. The cells are in an isotonic solution in the center.
9. Review Figure 1.2. Which solutions are hypotonic to the cytoplasm of the red blood cells, and which are hypertonic? How do you know?

Figure 1.2. Red blood cells along a concentration gradient.

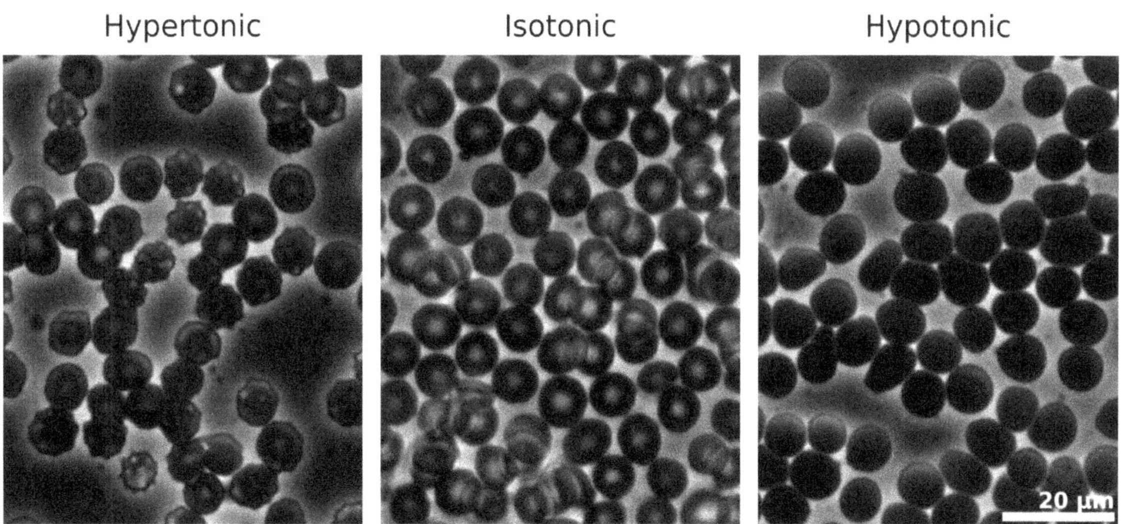

Name: _________________________ Lab Time: _______________ Due: _________________

2 | Enzymes Pre-Lab

Instructions

- ❏ Read the lab manual and then follow the instructions to complete the assignment.
- ❏ You may need your textbook and other resources to complete this assignment.
- ❏ Pre-labs must be completed before the start of the lab.

1. What is the function of the active site of an enzyme?

2. Can the shape of an active site be changed to inhibit or, perhaps, speed up a reaction? Explain your reasoning.

3. Complete Table 2.1 by briefly explaining the role of each of the following compounds in this lab.

Table 2.1. Explain the Roles of Each Compound

Compound	Role in lab
Amylase	
Lugol's Iodine	
Maltose	
Starch	

Complete Table 2.2 by stating the question that is being asked by each experiment we perform in each lab and providing a valid hypothesis.

Table 2.2. Provide a Hypothesis

Exercise	Question Being Asked by Experiment	Why do we do this experiment?
1		

Exercise	Question Being Asked by Experiment	Your Hypothesis
2A.		
2B.		

<table><tr><td># 2 | Enzymes</td><td>2</td></tr></table>

2 | Enzymes

By the end of this lab, you should be able to:
- ❑ Describe how changes in pH and temperature affect the function of an enzyme.
- ❑ Explain how changes in enzyme concentration impact the rate of a chemical reaction.
- ❑ Explain how enzymes function and why their structure is important.

The cytoplasm of the cell is packed with a variety of proteins, many of which function as enzymes. **Enzymes** are biological catalysts that function to speed up chemical reactions. These reactions either build complex molecules (anabolic) or break them down (catabolic). The reaction that takes place depends on the enzyme involved.

During a chemical reaction in the cell, an enzyme acts on a **substrate** (or substrates), creating a **product**. Substrates and products differ from each other in structure, but enzymes remain unchanged by the reaction.

Activity 1. Detecting Amalyse Activity

Amylase is an enzyme present in the saliva of animals that aids in the degradation of starch. **Starch** is a polysaccharide created by plants. When animals consume starch, it must be broken down to smaller mono- and disaccharides in order to be absorbed by the digestive system.

Amylase catalyzes a **hydrolytic** chemical reaction that breaks starch down into the disaccharide maltose, which contains two monomers of glucose attached to each other by a **glycosidic bond** (Figure 2.1). Other enzymes in the digestive system further break down maltose before its components are absorbed in the small intestine.

Figure 2.1. Hydrolytic reaction of amylase with starch.

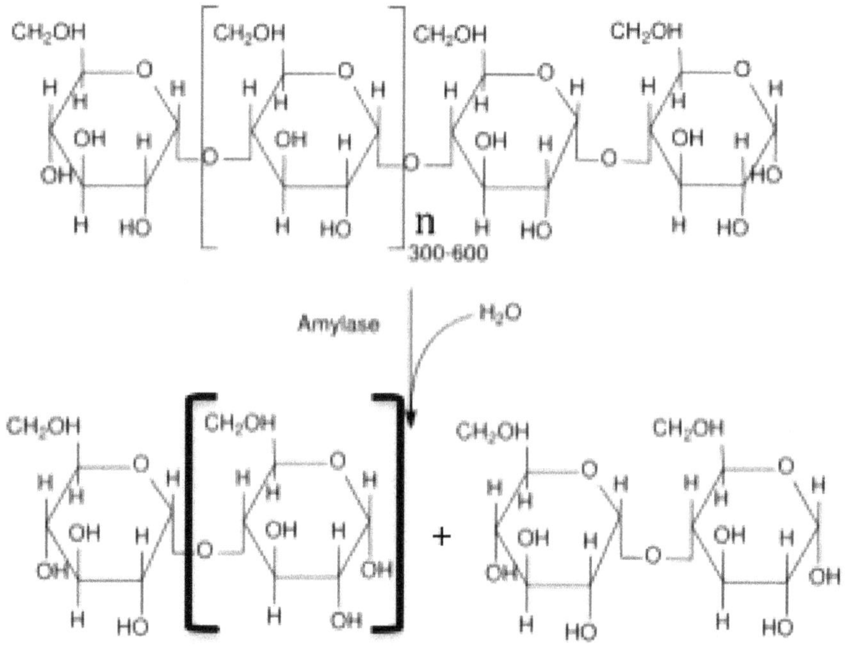

Procedure A. What Reacts with Lugol's Iodine?

The rate of a reaction can be determined by either the disappearance of substrate or the appearance of product. In the reaction we are examining in this lab, starch is the substrate and maltose is the product. We will use Lugol's iodine to measure this amylase activity, but first, we need to determine what it reacts with.

To observe the interaction of Lugol's iodine with starch and maltose, obtain the following:

- ❏ 1% maltose solution
- ❏ 1% starch solution
- ❏ 3 Test tubes
- ❏ Distilled water
- ❏ Dropper bottle of Lugol's iodine

Instructions

1. Label one test tube "starch," one test tube "maltose," and the other test tube "water."
2. Use a graduated pipette to add 2 mL of the appropriate solution into each tube.
3. Add 5 drops of Lugol's iodine to each tube.
4. Record your observations by answering the questions below in your lab notebook.
   a. Describe the appearance of Lugol's iodine diluted in water.
   b. What happens when you add Lugol's iodine to starch?
   c. What happens when you add Lugol's iodine to maltose?

Q1. Why did we add Lugol's iodine to water? Did that tube help you interpret your data?

Q2. What would happen to the color of solutions containing diluted starch mixed with Lugol's iodine? The addition of Lugol's Iodine to a solution denatures amylase so that it can no longer carry out the chemical reaction described earlier.

Q3. Explain how Lugol's iodine is used to examine the activity of the amylase enzyme.

Procedure B. General Assay for Amylase Activity

Figure 2.2. Experimental set up for amylase assay.

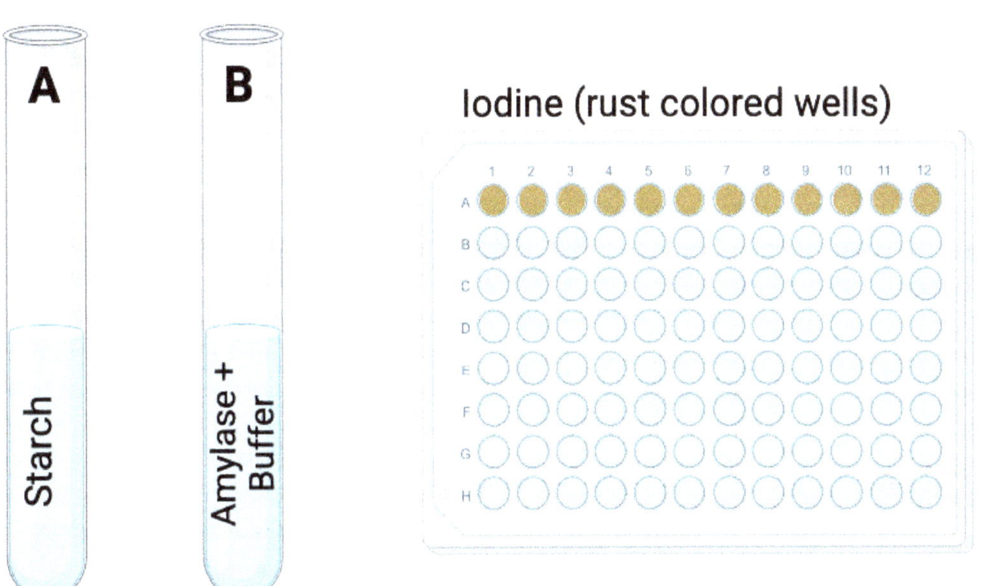

Review the experimental set up for amylase assay in Figure 2.2. Tube A contains 2 ml of starch, and Tube B contains 1 ml of amylase and 1 ml of buffer. Each well in the top row of a 96-well plate contains one drop of Lugol's Iodine solution.

Instructions

1. Obtain two clean test tubes and label them "A" and "B" using a wax pencil.
2. Add the following to each tube (See Figure 2.2):
   a. To Tube A, add 2 mL of starch.
   b. To Tube B, add 1 mL of 1% amylase and 1 mL of pH 6.8 buffer.
3. Allow the tubes to rest for 2 min (this allows the enzyme to acclimate to its environment.
4. While you are waiting, add 1 drop of Lugol's iodine to each compartment in one row of the test plate.
5. Place the test plate on a piece of white paper.

> A teammate should start the timer at the moment the contents of the tubes are combined (Time 0) in the next step.

6. After the acclimation period has ended, use a clean plastic pipette to transfer the contents of Tube B into Tube A. Mix the solution by sucking it up and down with the pipette.
7. After 10 seconds, transfer a drop of the reaction mix (from your test tube) to the first well on the plate. The contents of the well should turn a purplish-blue color. If not, get help.
8. Repeat every 10 seconds until no color change is observed (the solution in the well remains amber-orange).
9. Count the number of wells used before the color stopped changing.

Q4. How many wells did you add the reaction mixture to before you noted "no reaction?"

Q5. Each well represents 10 seconds. Based on that, how long did it take for the reaction to happen?

Q6. Why do we put amylase and starch in different test tubes?

Q7. What is the role of the buffer in this experiment?

Q8. What does the stop in color change you are looking for represent in terms of the chemical reaction we are observing?

Activity 2. Altering the Rate of Amylase Activity

If you have followed along so far, you understand that if we mix starch and amylase, the starch will be broken down over time. We can see this degradation by, at specific time intervals, adding Lugol's iodine to the enzyme mixture to both halt the chemical reaction and to observe how dark its coloring is. As the starch is degraded, the enzyme mixture should produce a less and less purple color when Lugol's iodine is added.

In this activity, we will take advantage of our ability to "see" amylase in action as we study how several factors alter its rate of action.

- ❑ Each group will be assigned to complete part A or B only.
- ❑ You are responsible for gathering the data from the other groups in your class.
- ❑ The results of replicate experiments should be averaged to create your final graphs for each experiment.

> For consistency, always add starch and water to the tube labeled A. Add amylase and buffer to the tube labeled B.

Instructions
1. Define the hypothesis you are testing in the space provided.
2. Design the experimental procedure.
3. All the teams assigned to pH will work together to design this experiment so that your results can be averaged together.

4. Obtain the bucket labeled pH from your instructor and examine its contents.

5. Write a step-by-step procedure to test your hypothesis on a white board.

6. Design a data table for your experiment on the top half of a piece of graph paper. Tables should include columns for data from the other groups testing this independent variable.

7. Record a prediction about what you think will happen in each experiment based on your hypothesis in the space provided.

> DO NOT START YET! Review your procedure and data table with your instructor. You must have your instructor's initials in your lab notebook before moving on. _________

8. After getting your instructor's approval, run the experiment.

9. Record the data as you gather it in your data table.

10. Complete your table by gathering data from other groups that ran the same experiment.

11. Graph the averaged data of all the groups running the experiment on the bottom half of the graph paper.
    a. Each student should create their own properly formatted graph.
    b. All graphs should be made by hand on graph paper.

12. Obtain the data table from the groups that tested the alternate independent variable and create a graph as you did for your experiment.

13. Create separate scans of the pH table and graph, and the temperature table and graph. You may be asked to upload a scan of either on the lab quiz.

Q9. What variable was your group assigned to test?

Q10. What factors affect the rate of amylase activity?

Q11. Hypothesis (what is your answer to the question above?):

Procedure A. The Effect of pH on Enzyme Activity

pH has a significant impact on the three-dimensional structure of a protein. Hydrogen (H) bonds and ionic bonds determine much of the secondary, tertiary, and quaternary structures of proteins.

When these bonds/attractions break, the protein unfolds a bit and changes shape, a process known as denaturation. Since shape determines enzyme structure, a change in structure impacts its function.

Q12. **Prediction:** What do you think the results of the experiment will be?

> Be sure to allow time for the enzyme to acclimate to the different pH environments. If you have seen no color change after 7 minutes, you can safely conclude that the enzyme does not work (has 0 activity) at that pH.

Q13. Why does the enzyme need to acclimate before you start the reaction?

Q14. What is the pH optimum of amylase? Explain how you know.

Q15. What is happening to the enzyme at the molecular level in environments that are lower than the pH optimum?

Q16. What is happening to the enzyme at the molecular level in environments that are higher than the pH optimum?

Q17. Did your results support your hypothesis? Why or why not? Be sure to use your results to answer this question.

Procedure B. The Effect of Temperature on Enzyme Activity

The activity of enzymes is also impacted by temperature. Temperature is a measure of the kinetic energy of molecules. That energy can have a few different impacts on enzyme activity. The movement of molecules can determine how often an enzyme finds its substrate and high temperatures can break chemical bonds causing denaturation.

Q18. **Prediction:** What do you think the results of the experiment will be?

Tips!

- ❑ Use 30-second intervals for testing during this experiment.
- ❑ If you have seen no color change after 10 minutes, you can safely conclude that the enzyme does not work (has 0 activity) at that temperature.
- ❑ The solutions used in this experiment must be allowed to "get to temperature" before you start the experiment.
- ❑ The reaction mixture for these experiments should be kept at their assigned temperature throughout the testing period.

Analyzing Your Results:

Q19. What is the temperature optimum of amylase? Explain how you know.

Q20. What is happening to the enzyme at the molecular level in environments that are lower than the temperature optimum?

Q21. What is happening to the enzyme at the molecular level in environments that are higher than the temperature optimum?

Q22. Did your results support your hypothesis? Why or why not? Be sure to use your results to answer this question.

Q23. In this experiment, you placed one set of tubes in ice. What would happen if you moved the reaction tube for this sample to the 37°C incubator and then retested?

Q24. Would the same thing happen if you moved the reaction tube in the 80°C incubator to the 37°C incubator? Why or why not?

2 Applying What You've Learned

A selection of these questions will appear on your lab quiz. Lab quizzes are open book. Type up answers to these questions in advance so that you can copy and paste the answers into the quiz.

Figure 2.3. These graphs show the rate of enzyme activity of three enzymes (enzymes A–C) under different pH and temperature conditions.

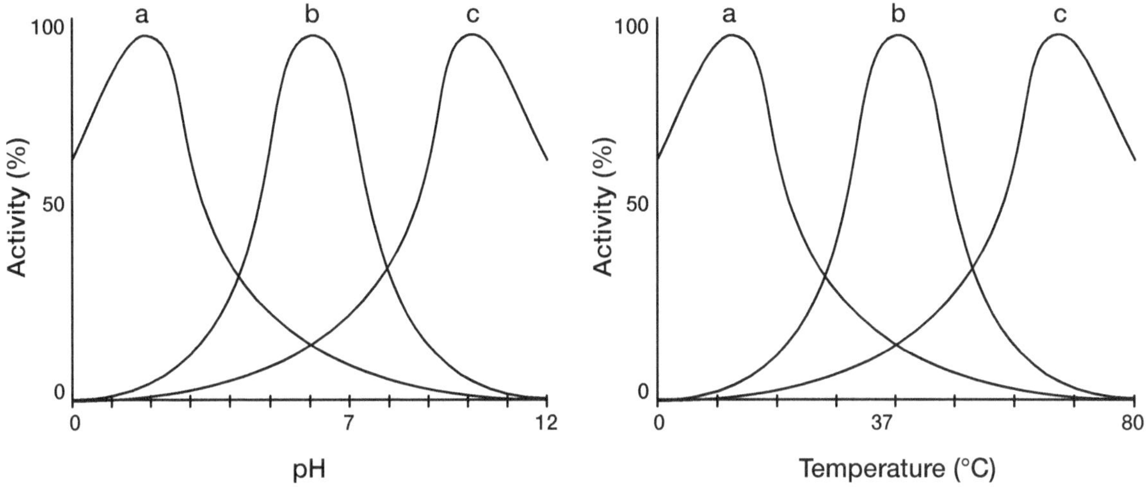

1. Which of the enzymes in Figure 2.3 is most likely to be involved in digestion in the human stomach?
2. Which of the enzymes in Figure 2.3 is most likely to be found in the blood of a fish that lives in the Arctic Ocean?
3. Which of the enzymes in Figure 2.3 is most likely to be found in a bacterium that lives in a hot spring in Yellowstone National Park?

One of these questions will appear on the lab quiz. Note that there are several aspects to these processes, and we are focusing only on those that relate to this lab. Students who look up the answer rarely include the information I am looking for here. Think it through on your own!

4. When we eat meat, we usually cook it first. As it cooks, the texture of the meat changes. Cooked meat is easier to chew and digest. Use what you have learned in this lab to explain why.
5. Ceviche is a Peruvian dish that consists of raw fish marinated in citrus juice and seasonings. The process of making ceviche "cooks" fish without heat. How is this possible?

<table>
<tr><td>**3**</td><td># Restriction Enzymes & Electrophoresis Pre-Lab</td></tr>
</table>

Instructions
- ❑ Read the lab manual and then follow the instructions to complete the assignment.
- ❑ You may need your textbook and other resources to complete this assignment.
- ❑ Pre-labs must be completed before the start of the lab.

Use QR code 3.1 to watch the Amoeba Sisters DNA Electrophoresis video and the answer the following questions:

QR Code 3.1. Amoeba Sisters DNA Electrophresis.
https://www.youtube.com/watch?v=ZDZUAleWX78

1. What is the charge of a molecule of DNA?

2. During agarose gel electrophoresis of DNA:
    a. What causes the DNA to move through the gel? _______________
    b. Based on what characteristic of the DNA are pieces of DNA separated? _________

3. When looking at an agarose gel of DNA, where would the biggest pieces of DNA be located? The smallest?

> Use your textbook or other resource to find information about bacteriophage lambda (λ). We will be using DNA from this phage in our laboratory.

4. What features of bacteriophage l make it a good subject for biological studies?

Activity 1. Working with Restriction Maps

Examine the following restriction map of bacteriophage λ. A **restriction map** shows all the locations where a specific restriction enzyme(s) would recognize and cut DNA, as seen in Figure 3.1.

The vertical lines on the ends indicate the number of base pairs at the start and end of the bacteriophage's DNA molecule. The vertical lines between the ends indicate the location of all the cuts that would be made if bacteriophage λ DNA were mixed with the restriction enzyme HindIII. The numbers associated with the lines indicate the position (in base pair number) where HindIII cuts the DNA.

Figure 3.1. A restriction map showing enzyme cutting DNA.

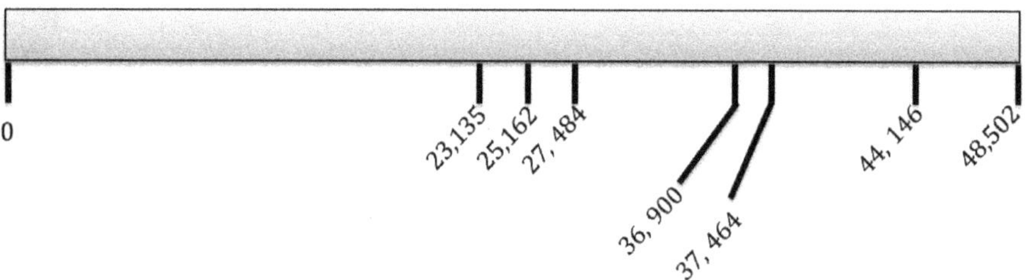

Subtracting the numbers will give you the number of base pairs (the size of the DNA fragment) formed by cutting in those two spots. When you mix DNA with an enzyme, assume that every restriction site is cut every time.

5. How many pieces of DNA will be created by cutting bacteriophage l DNA with HindIII?

6. List the sizes of all of the DNA fragments created by this digest.

Instructions

1. Label the positive and negative ends of the diagram of the electrophoresis gel.
2. Add bars (rectangles) to the "Digested DNA" column to predict what the agarose gel of bacteriophage λ DNA treated with BamHIII would look like after electrophoresis.
3. Label each band in the gel with the size of the DNA fragment it corresponds to.

Use the Marker DNA column as an example of what to do. Bands in the Digested DNA column should be arranged relative to those in the Marker DNA column (in regard to size). Use the position of bands in the Marker DNA column to guide the position of bands in the Digested DNA column.

Figure 3.2. A diagram of an electrophoresis gel containing two samples.

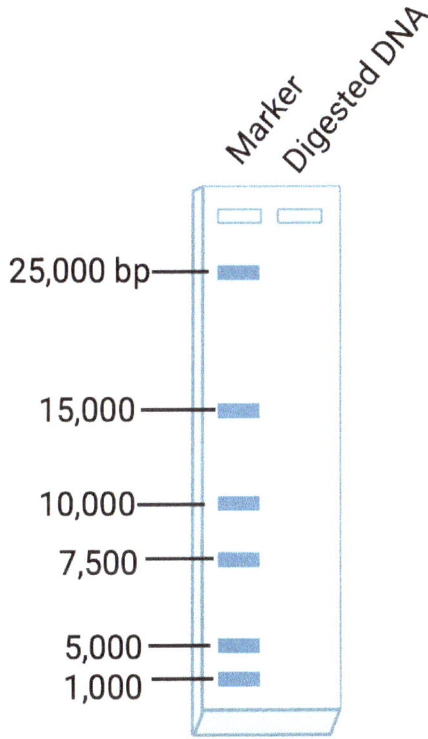

Figure 3.2 includes a diagram of an electrophoresis gel containing two samples. The rectangles labeled "Marker DNA" and "Digested DNA" indicate the wells into which the DNA was loaded. All the pieces of DNA in the "Marker DNA" column are of known size, and the numbers within the rectangles represent the size of the DNA in that "band."

3 | Restriction Enzymes & Electrophoresis

By the end of this lab, you should be able to:
- ❑ Describe the function of restriction enzymes.
- ❑ Perform and explain basic techniques in molecular biology (restriction enzyme digest and agarose gel electrophoresis.
- ❑ Use a restriction map to predict the outcome of a restriction enzyme digest.

Molecular biology is a field of study that focuses on how the organic compounds that build cells function. The manipulation of DNA is one way scientists gain an understanding of how cells work. **DNA splicing** is the process of cutting out and reassembling pieces of DNA to create recombinant DNA that can be inserted into an organism to alter its function.

In this lab, we will demonstrate how special enzymes extracted from bacteria can be used to cut DNA at specific DNA sequences, the first step in DNA splicing.

Restriction enzymes evolved in bacteria as a form of protection against **bacteriophages**, viruses that infect bacteria. When bacteriophages enter a bacterial cell, they insert their DNA into the bacterial chromosome and utilize the cell's machinery to transcribe and translate their genes, ultimately building more virus particles.

Restriction enzymes evolved to recognize specific DNA sequences (**restriction sites**) and cut the covalent bond between DNA nucleotides at specific locations in that sequence. Once the covalent bond is broken, the weak hydrogen bonds that hold base pairs together separate, creating fragments of DNA. Bacteria with restriction enzymes can break down viral DNA before it is inserted into the bacterial chromosome.

Activity 1. Restriction Enzyme Digest of Bacteriophage λ

In this lab, we will work with bacteriophage λ. The DNA of this virus includes approximately 48,000 base pairs that encode numerous proteins. Several restriction sites, each corresponding to a different enzyme, are located within the bacteriophage λ sequence. In this lab, we focus on three restriction enzymes:

- ❑ EcoR1 (restriction site GAATTC)
- ❑ HindIII (restriction site AAGCTT)
- ❑ Pst1 (restriction site CTGCAG)

Note that these restriction sites are all **palindromic** (i.e., they read the same backward and forwards). Due to standard convention, we only list one strand of the DNA that is recognized by the enzyme.

If you write the DNA sequence of the other nucleotide strand of the DNA helix underneath, you will find that it reads the same sequence (from 5' to 3').

All restriction sites are palindromic, and the enzyme always cuts the covalent bond between the same two nucleotides on both strands.

For example, EcoRI cuts between the G and the A on both sides of its restriction site, as seen in Figure 3.3.

Figure 3.3. EcoRI cuts between the G and the A on both sides of its restriction site.

This restriction enzyme will always cut the restriction site in exactly this way, regardless of the species from which the DNA originates. Since every restriction enzyme cuts at different restriction sites, the same piece of DNA will give different results for each enzyme.

Once the DNA is cut into smaller fragments by restriction enzymes, agarose gel electrophoresis is used to separate the fragments by size, and stains are used to visualize the DNA.

Your instructor will demonstrate the use of a micropipette. If you have questions, ask your instructor for assistance.

1. Obtain a set of solutions and samples from your instructor. Familiarize yourself with the contents (Table 3.1).

Table 3.1. Contents of Tubes Used in Restriction Enzyme Digestion

Label	Contents
M	Marker (contains pre-digested DNA fragments of known size)
λ	Stock solution of λ DNA
RB	6 µl restriction buffer
P	5 µl restriction buffer + 1 µl *Pst* I
E	5 µl restriction buffer + 1 µl *EcoR*1
H	5 µl restriction buffer + 1 µl *Hind*III

2. Move the M tube and the λ tube out of the foam float, but keep them on ice.
3. Use the micropipette to add 4 µL of λ DNA to each of the colored tubes (RB, P, E, and H). Use a new tip for each tube.
4. Close the tube and gently flick the tube with your finger to mix.
5. Place the colored tubes directly across from each other in a microcentrifuge. If your tubes are very small, you will need to nest them in a larger tube. Spin the tubes for 2-5 sec (one pulse) to gather all the solution at the bottom of the tube.
6. Place the colored tubes in the foam float and incubate for 30 minutes in a 37°C water bath.
7. At the end of the incubation period, place your tubes on ice.

Q1. What is the function of a DNA marker?

Q2. You will add DNA from the λ stock solution to the RB, P, E, and H tubes. What is the function of the RB tube?

Activity 2. Agarose Gel Electrophoresis

Once the DNA has been digested, the resulting fragments must be separated to visualize the "pattern" created by the DNA fragments. Molecular biologists use the process of agarose gel electrophoresis to separate these fragments based on their size.

Recall that DNA is packed with negatively charged phosphate groups. These phosphates give the molecule an overall negative charge. As a result, DNA will move toward the positive end of an electrical field.

To separate DNA fragments by size as they migrate through this electrical field, we place the DNA mixture in the well of an agarose gel. Agarose is a polymer extracted from seaweed that creates a molecular matrix like a sieve. Smaller DNA molecules move more quickly through the gel than larger DNA molecules.

As a result, the rate at which the molecule moves through the gel is proportional to its size, and fragments of DNA that are the same size will travel the same distance.

When we perform this technique, we use not just one copy of the DNA, but many copies. The restriction enzymes cut all these copies in the same places, producing many copies of each size DNA fragment.

Since all fragments of the same size travel together through the gel, they will end up concentrated in a specific position called a band. When the gel is stained with a dye that binds DNA, each differently sized fragment of DNA will be represented by a different band on the gel.

Procedure A. Practice Loading a Gel

Your instructor will provide a demonstration for each group.

Instructions

1. Cover your practice gel with a layer of water. The water should be deep enough that its surface of the water is mirror smooth.
2. Use a micro-pipette to obtain 10 µl of practice dye (PD).
3. Slowly drop the dye into a well as described by your instructor.
4. Everyone in your group should load at least one well on the practice gel.
5. After everyone has practiced, rinse out your practice and return it to the group bucket.

Procedure B. Add Loading Dye

Instructions

1. Obtain a tube of loading dye (LD) from your instructor.
2. Add 2µl of LD to each of your colored tubes and to the M tube.
3. Return your tubes to the foam float and incubate in a 65°C water bath for 5 min.
4. Return your tubes to ice while you set up the gel.

Procedure C. Gel Loading

Instructions

1. Place your agarose gel in the electrophoresis apparatus.
   a. Be sure to orient the wells toward the appropriately charged end of the electrical field.
2. Place the electrophoresis apparatus near the power source that you will be sharing with the other team at your lab bench.
   a. Be sure that the cables on your apparatus reach the power source, as you cannot move the gel once the samples have been loaded.
3. Add enough 1X TAE to cover the surface of your agarose gel. The solution should fill the wells on both sides of the gel and cover the gel. The surface of the solution should be smooth.

Check with your instructor before continuing.

4. Place your samples in the micro centrifuge directly across from each other.
5. Use the pulse button to spin them briefly (10 seconds).
6. Make a quick sketch of your gel in your lab notebook.
7. Using a different tip for each sample, load 10 µl of each sample into a different well in the gel.
8. To help with orientation, load the M sample into the first lane, skip a lane, and then place your other samples in order in the remaining lanes.

As you load the gel, indicate which sample went into which lane in your lab notebook.

9. Carefully slide the cover onto the electrophoresis apparatus and plug your gel into the power source.
10. When the other team using your power source has attached their gel, turn on the power to high.

If everything is set up correctly, you should see tiny bubbles streaming up the short sides of the apparatus.

11. Allow the gel to run until the blue band of stain has moved 3-5 mm away from the wells.

Activity 3. Predicting Your Results

We know the DNA sequence of our bacteriophage, and the specific restriction sites are recognized by the restriction enzymes we are using. As a result, we can make a prediction of what our gels will look like.

To do this, we need to calculate the size of the DNA fragments we would expect to see if each of the enzymes we used cut the DNA at every possible restriction site. Then, using our understanding of agarose gel electrophoresis to sketch an image of what the gel will look like.

1. **Hypothesis:** Restriction enzymes cut DNA in predictable places.
2. **Procedure:**
3. In the space provided, predict the sizes of DNA fragments that would be produced by mixing bacteriophage DNA with EcoRI (Figure 2) and HindIII (separately).

> Note: You already did one of these in your pre-lab.

4. Use your calculations and the pre-calculated DNA fragment lengths from Pst I in Table 3.2 to complete the agarose gel in Figure 3.4 that should be produced by your three digests.

Figure 3.4. Known size (base pairs) of marker DNA fragments

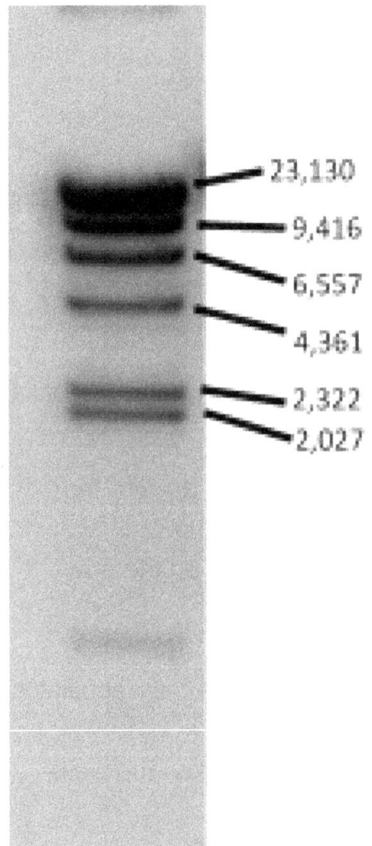

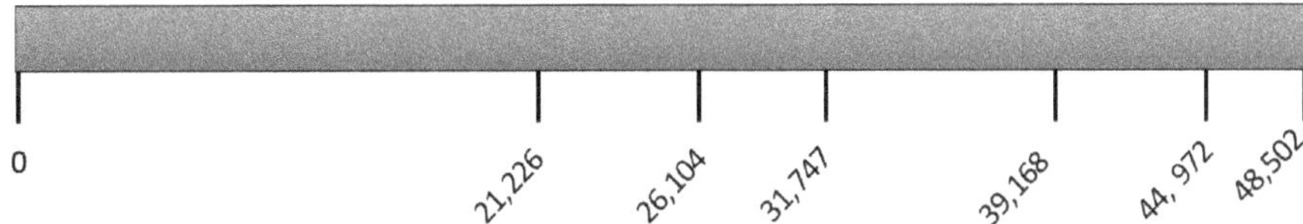

Table 3.2 includes the sizes of the DNA fragments in the DNA marker. The number associated with each vertical line indicates the base pair number where EcoR1 cuts bacteriophage DNA. Base pair number runs from 0 bp on the left to 48,502 bp on the right.

Table 3.2. Size of Major DNA Fragments Produced by Pst1 Digest of Bacteriophage

Major DNA fragments produced by digestion of bacteriophage (bp) with Pst I	
11497	2459
4749	1986
2838	1700

Pst I has 29 restriction sites in bacteriophage DNA. With so many restriction sites, many of the DNA fragments overlap in size or are too small to see with our equipment. The size (bp) of the most easily visualized fragments is listed.

Q3. Which lane contains your DNA marker? What is the role of this lane?

Q4. Where on the gel will you find the largest DNA fragments? Smallest?

Activity 4. Analyzing Your Results

You can learn a great deal by simply examining your gel, but accurate results require further analysis. In this activity, you will examine the results of your restriction digest in two ways:
- ❑ Estimation
- ❑ Using a standard curve on semi-log graph paper

Procedure A. Documenting Your Results

Instructions

1. Place the completed gel on the blue light box to see the DNA bands.
2. Align a ruler next to the gel for measuring purposes
3. Make a sketch of your gel in the space provided. Your instructor will also provide a photograph.
4. Label the lanes according to the sample they contain.
5. Label the size of each DNA band in the DNA marker (see Table 3.3).
6. Sketch or photograph the gel produced by agarose gel electrophoresis in Figure 3.6 at the end of Lab 3.

Procedure B. Estimating the Size of DNA Fragments

Sometimes, you just want to do a quick check of your results. Since DNA fragments of the same length move the same distance in an agarose gel, you can quickly assess your gel using the relative position of bands of known sizes in the DNA marker lane.

Instructions

1. Use a straight edge to align the bottom of an unknown band to the marker lane.
2. Use the nearest marker band and your eyes to estimate the size of the DNA in that band based on its position in relation to the known size of the bands in the DNA marker lane.

Remember that DNA of the same size will travel the same distance in the gel regardless of which lane it is in.

3. Repeat for all unknown bands.
4. Record your estimates in the appropriate columns of Table 3.3.

Table 3.3. Estimation of Bacteriophage λ Restriction Digest Fragments (not all cells will be used)

	DNA marker (M)	No enzyme (RB)	*Pst*I (P)	*Eco*R1 (E)	*Hind*III (H)
	Actual size (base pairs)	Estimated size (base pairs)	Estimated size (base pairs)	Estimated size (base pairs)	Estimated size (base pairs)
Band 1	23,130				
Band 2	9,416				
Band 3	6,557				
Band 4	4,361				
Band 5	2,322				
Band 6	2,027				

Procedure C. Using a Standard Curve to Calculate the Size of DNA Fragments

Instructions

Create a standard curve.

1. In the DNA marker lane, use a ruler to measure the distance from the bottom of the well to the bottom of each band
2. Arrange your measurements from largest fragment to smallest fragment and record in Table 3.4.
3. Use the measurements from the DNA marker lane to plot each band on the semi-log graph provided at the end of this lab (Figure 3.6).
4. Create a "best fit" line on the graph.

Use the curve you created.

5. Use the standard curve to determine the size of each of the other bands in every lane (not the "M" lane) of your gel.
6. Use a ruler to determine the distance traveled from the well to the bottom of each band in each lane.
7. Record your data in the appropriate columns (distance travelled, mm) of Table 3.4.

8. Find the distance traveled by a band on the X-axis of the standard curve.

9. Use a straight edge to travel from the X-axis to the standard curve and then from the curve to the Y-axis. The value on the y-axis represents the size of the DNA in that band.

10. Record the size of that band in Table 3.4.

Table 3.4. Estimation of Bacteriophage λ Restriction Digest Fragments Based on Standard Curve (not all cells will be used, but you should use the same number as you did in Table 3.3)

DNA Marker (M)		No Enzyme (RB)		*PstI* (P)		*EcoR1* (E)		*HindIII* (H)	
Known size (bp)	Distance Traveled (mm)	Distance traveled (mm)	Calculated Size (bp)	Distance traveled (mm)	Calculated Size (bp)	Distance traveled (mm)	Calculated Size (bp)	Distance traveled (mm)	Calculated Size (bp)
23,130									
9,416									
6,557									
4,361									
2,322									
2,027									

Q5. Revisit your hypothesis and prediction (compare your actual gel to the sketch you made previously). Does your data support your hypothesis? Record your thoughts and conclusions below.

3 | Applying What You've Learned

A selection of these questions will appear on your lab quiz. Lab quizzes are open book. Type up answers to these questions in advance so that you can copy and paste the answers into the quiz.

1. Use your data to explain how you know that each restriction enzyme cuts the bacterio-phage DNA at different locations.
2. The DNA sequences recognized by restriction enzymes must be palindromic, but they can include any number of base pairs. Is a restriction enzyme with a long restriction site (many base pairs) more likely to cut a strand of DNA than a restriction enzyme with a short restriction site?
3. Is a restriction enzyme digest more or less likely to produce the same size DNA fragments in two individuals that are closely related or two individuals that are not related?
4. You used two techniques to determine the size of the DNA fragments produced by your restriction digests: estimation and a semi-log graph. Which of these methods is more accurate?

Figure 3.6. Sketch or photograph the gel produced by agarose gel electrophoresis.

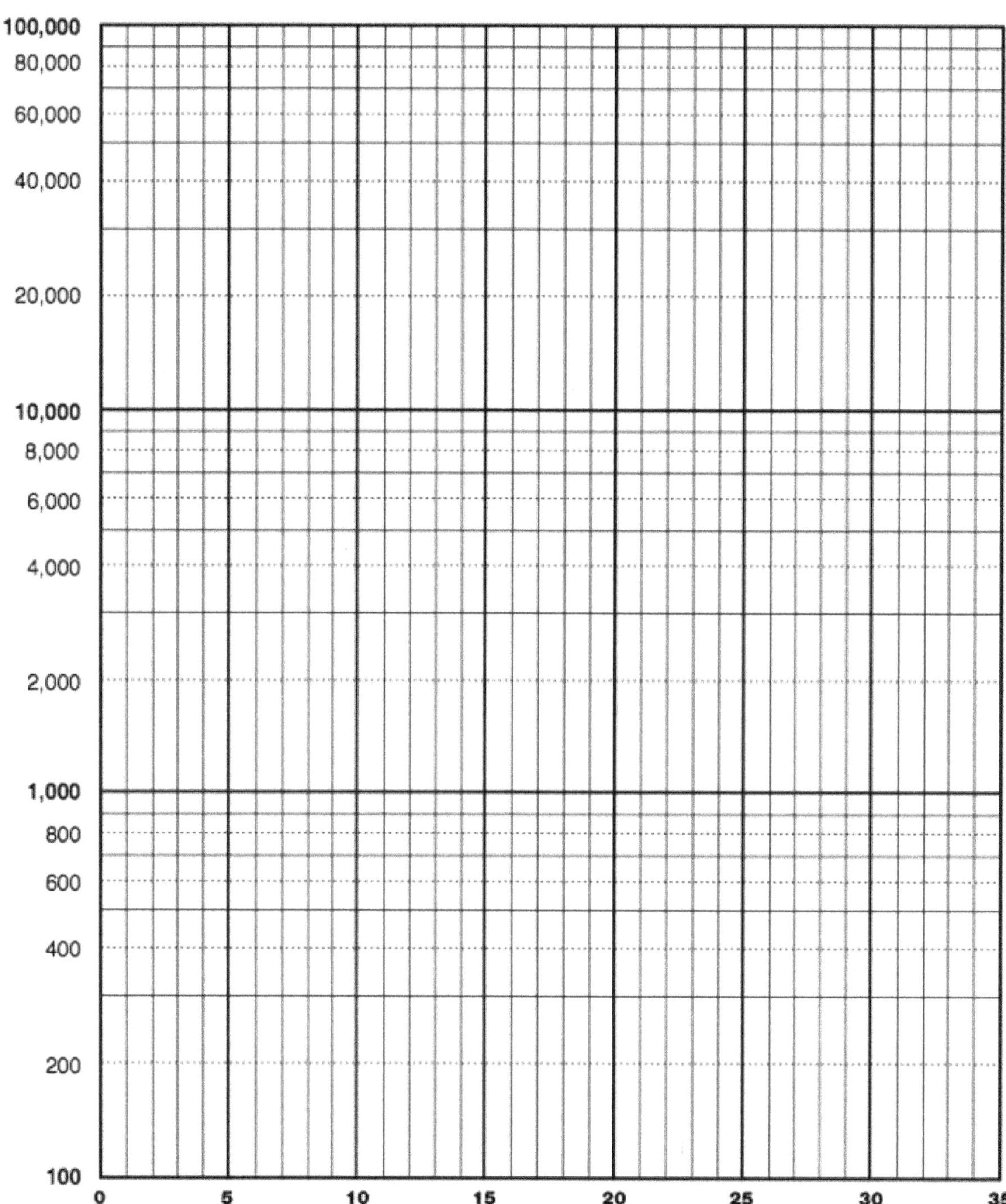

4 | Cells and Microscopy Pre-Lab

Instructions

- ❑ Read the lab manual and then follow the instructions to complete the assignment.
- ❑ You may need your textbook and other resources to complete this assignment.
- ❑ Pre-labs must be completed before the start of the lab.

To prepare for this week's lab, review Appendix A of the lab manual.

1. Complete Table 4.1 by adding the magnification of each object on the class microscope (A.1) and the diameter of the field of view.

Table 4.1. Diameter of Microscope Field of View at Different Magnifications

Microscope Objective	Field of View (μm)

Figure 4.1 is an image of tardigrades taken through the 4X objective of the class microscope.

Figure 4.1. Tardigrades through a 4x microscope objective.

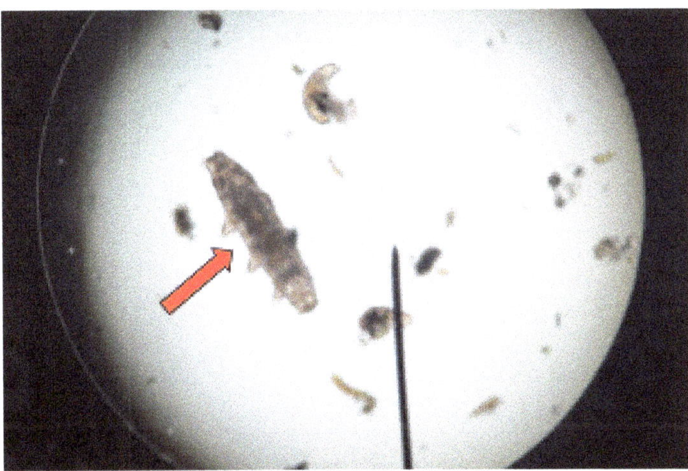

2. Estimate the length of the tardigrade indicated by the red arrow using the information you provided in Table 4.1.

This process is reviewed in Appendix A. You can also use QR code 4.1 to access a tutorial.

https://depts.washington.edu/vurchin/index.php?view=micmeas

3. Complete Table 4.2 by indicating the smallest size of an object that can be seen with your eyes, light, and electron microscopes.

Table 4.2. Recording Data Using Types of Observation

Type of Observation	Size Smallest Object That Can Be Visualized (include units)
Naked eye	
Light microscope	
Electron microscope	

4. Use the Internet to find the approximate size of the organelles in Table 4.3.

5. Complete Table 4.3 as instructed in the appropriate columns.

> If you've been searching for the answer for more than 5 minutes and can't find it, just write "not found" in the box.

Table 4.3. Approximate Size of Organelles

Organelle	Approximate Diameter on longest axis (include units)	Type of microscope you should use to see
Chloroplast		
Parts of the cytoskeleton		
Endoplasmic Reticulum		
Golgi Complex		
Mitochondria		
Nucleus		
Ribosome		

You will be required to create scaled drawings during this lab. This process is described in Appendix A and in the tutorial linked to QR Code 4.2.

QR Code 4.2. Tutorial for scaled drawings.
http://go.chemeketa.edu/Lab4Prelab

4   # Cells & Microscopy

By the end of this lab, you should be able to:

- ☐ Describe the basic structure of a eukaryotic cell.
- ☐ Differentiate between several types of microscopes in regard to their function and use.
- ☐ Explain several ways that specimens can be prepared for microscopy.
- ☐ Use a microscope (or a micron bar in a micrograph) to estimate the size of a cell or organelle.

Activity 1. Introduction to Microscopy

Microscopy is the use of a microscope to study a specimen (living or not). Modern microscopy relies on the use of several different types of microscopes. When a scientist decides to use microscopy to test a hypothesis, he or she must carefully consider what information needs to be gathered to determine what type of microscope to use.

Procedure A. Reviewing Different Types of Microscopy

Use these keywords to complete Table 4.1 as your instructor describes the different types of microscopy.

Staining
- ☐ Chemically colored
- ☐ Fluorophore
- ☐ Heavy metal
- ☐ None

Type of illumination
- ☐ Transmitted, reflected, or excitatory
- ☐ White light, specific wavelengths of light, or an electron beam

Type of specimen
- ☐ Living
- ☐ Preserved (whole mount or sectioned)

Type of Microscope	Type of illumination	Type of specimen	Size of specimen	Staining
Compound Light Microscope				
Stereomicroscope				
Fluorescence Microscope				
Transmission Electron Microscope				
Scanning Electron Microscope				

Activity 2. Using the Compound Light Microscope

If you are new to microscopes, you should spend some time reviewing the parts of the microscope and their function. You may find the virtual microscope in QR Code 4.3 to be a helpful tool.

QR Code 4.3. Virtual microscope.

https://depts.washington.edu/vurchin/index.php?view=micmeas

Appendix A contains a labeled diagram of the microscopes used in this class, along with related procedures.

Activity 3. What Can You See with a Compound Light Microscope?

Eukaryotic cells contain a variety of organelles that perform specific functions in the cell. Particularly of interest are the membrane-bound organelles, some of which are large enough to see with a compound light microscope. In this activity, we will explore several cell types to learn how to identify these different organelles.

How do you figure out which organelle is which?
- ❑ Size is the most important indicator of organelle type.
- ❑ Color may be a good clue if your specimen is alive.
- ❑ You may need to infer that an organelle is present due to the arrangement of other structures in the cell rather than actually seeing its boundaries.

Procedure A. Drawing and Identifying Specimens

Instructions
1. Divide a page of graph paper into six sections. There should be three sections for sketches and three for accompanying observations.
2. Sketches should fill one of the sections. Plan to add as much detail as possible to the sketch. The bigger your sketch, the easier that will be.
3. Each sketch should include a scale bar (see Appendix A).
4. Each sketch should be associated with 3-5 observations of the specimen (located in the ⅙ of the page next to the sketch).

For each of the following specimens
5. Prepare a slide as directed.
6. Draw a sketch of a single cell from that specimen with as much detail as possible in your lab notebook.
7. Identify and label as many structures associated with the cell as possible.

Specimens

Elodea
8. Create a wet mount of a single leaf.

Onion Epidermal Cells
9. Obtain a single leaf from an onion.

10. Use your fingernail to gently scratch the top (innermost layer) of the leaf to separate the filmy epidermal tissue.
11. Create a wet mount from a small section of this filmy tissue. Lay the tissue as flat as possible on the slide to reduce air bubbles and folds.
12. Examine the wet mount with your compound microscope.
13. After examining the specimen, stain your specimen with Iodine as it sits on the microscope (see Appendix A for procedure).

Cheek Cells

14. Create a wet mount using 0.9% NaCl rather than water (water will cause the cells to burst).
15. Use the larger, flat end of a toothpick to gently scrape the inside of your cheek.
16. Twirl the toothpick in the drop of NaCl until the cells are released from the toothpick.
17. Place your toothpick in the appropriate disposal container.
18. Immediately mount a coverslip over your cells. Don't let your specimen dry out!
19. Examine the wet mount with your compound microscope.
20. After examining the specimen, stain your specimen with methylene blue as it sits on the microscope (see Appendix A for procedure).

Liver cells

21. Examine prepared slides.

> Note: To understand this material, you must look at the slides in the order described.

❑ Slide 1 (mammalian liver tissue, standard staining for contrast only): Check with your instructor to make sure you have really found a cell before moving on to the next step.

❑ Slide 2 (Amphiuma liver, specially stained for mitochondria): Check with your instructor to make sure you have correctly identified the mitochondria before moving on.

Q1. These four specimens contain all membrane-bound organelles (and other cell structures) that are easily visible with the compound light microscope. Based on this information and your observations, which cell structures are visible with a compound light microscope?

Q2. In multicellular organisms, many cells are organized into tissues. When looking at a prepared slide of such tissues, what features can you look for to identify a single cell and its boundaries?

Q3. We used iodine to stain the onion cells and methylene blue to stain the cheek cells. How did the addition of these stains affect your image of the specimen?

Q4. Methylene blue stains a specific organic compound. Based on your observations of the stained cheek cell, what organic compound does methylene blue attach to?

Activity 4. What Can You See with Other Types of Microscopy?

The light microscope enables us to examine certain structures within the cell. Special stains and dyes can be added to highlight certain aspects of the cells, but overall, scientists are limited in what they can see with a compound light microscope. A variety of microscopes are used to observe cells at different levels of detail. Although we cannot use these types of microscopy in this class, we can explore the types of images they produce.

Instructions
1. Examine the transmission electron micrographs provided (labeled A-F).
   a. Use what you have learned about the size and structure of eukaryotic organelles to identify each organelle.

    b. Record your identification and a scientific explanation of why you chose this organelle in your lab notebook.
2. Examine the scanning electron microscope images provided.
    a. Try to identify the specimen.
    b. The answers are on the back!
3. Examine the fluorescence micrographs (A-C) and explanations provided by your instructor.
    a. In your lab notebook, indicate what color (fluorophore) represents each part of the cytoskeleton and explain your reasoning.

Q5. How does the electron microscope differ from a light microscope in terms of function and use?

Q6. What are the advantages of using fluorescence microscopy instead of light microscopy to study a specimen?

4 | **Applying What You've Learned**

A selection of these questions will appear on your lab quiz. Lab quizzes are open book. Type up answers to these questions in advance so that you can copy and paste the answers into the quiz.

1. Consider the organelles you can see with both light and transmission electron microscopy. Describe the different types of information you can get from each form of microscopy.
2. Consider the size of mitochondria and chloroplasts. Do these sizes align with our hypothesis about the evolution of these organelles? *Hint: You may need to conduct some research to determine the size of the largest bacteria.*
3. In a few of these exercises, we applied chemicals (dyes) to stain the specimen. Provide two or more reasons why microscopists would use dyes and stains on their specimens.
4. One of these questions WILL be on the lab quiz:
   a. You are asked to determine how a particular drug affects the cytoskeleton of a cell. Design an experiment that utilizes one of the microscopy types that we discussed, which will allow you to answer this question. Be sure to include your reasoning as part of the answer.
   b. You are asked to determine how the overall structure of the endoplasmic reticulum changes when a cell is exposed to a particular hormone. Design an experiment that utilizes one of the microscopy types that we discussed, which will allow you to answer this question. Be sure to include your reasoning as part of the answer.
   c. You are asked to examine how individuals of a species of protist interact under different environmental conditions. Design an experiment that utilizes one of the microscopy types that we discussed, which will allow you to answer this question. Be sure to include your reasoning as part of the answer.

5 | Mitosis & the Cell Cycle Pre-Lab

Instructions

- ❑ Read the lab manual and then follow the instructions to complete the assignment.
- ❑ You may need your textbook and other resources to complete this assignment.
- ❑ Pre-labs must be completed before the start of the lab.

1. In this lab, we use prepared slides of onion root tips to "see" mitosis. What makes this specimen useful for observing mitosis?

2. Sketch the indicated chromosome(s) in Figure 5.1.

Figure 5.1. Chromosome sketches

A duplicated chromosone	A homologous pair of duplicated chromosomes	A pair of unduplicated, non-homologous chromosomes

3. Read Exercise 3 ("Which Stage of Mitosis is the Longest?") in this week's lab handout.

4. Create a hypothesis as to which stage of mitosis you think will be the longest.

5. Write a scientific explanation in support of that hypothesis.

5 Mitosis & the Cell Cycle

By the end of this lab, you should be able to:
- ❏ Describe cellular and chromosomal changes that occur during the cell cycle.
- ❏ Differentiate between cell division and cytokinesis.
- ❏ Explain how duplicated chromosomes are separated during cell division.
- ❏ Identify the stages of cells undergoing mitosis in prepared slides.

The **cell cycle** is the series of events that occur from the time a cell is formed until it creates new cells. Cells spend most of this cycle in a stage known as interphase. During interphase, the cell grows and performs the tasks assigned to it during development. Should the cell determine that it will undergo cell division, it replicates its DNA and begins to produce proteins involved in the cell division process.

Cell division refers to the steps involved in separating the copies of DNA produced by DNA replication at the end of interphase. Cells can undergo two types of cell division: mitosis and meiosis. Cell division is followed by cytokinesis, the physical separation of one cell into two.

In this lab, we will model the processes of mitosis using Play-Doh and explore the structure of chromosomes during each stage of cell division in both plant and animal cells.

Activity 1. Modeling the Cell Cycle & Mitosis

We discuss mitosis in terms of stages, but it is a dynamic process (flowing) rather than a step-by-step process in the cell. We just named stages to help us understand. The line between one stage and the next is often gray. In this activity, you will make model chromosomes and move them around to simulate what happens during mitosis.

Procedure A. Using the Modeling Clay

This procedure takes you through the steps of the cell cycle, including mitosis, using chromosomes made of clay as a model. Your model cell will contain a total of 3 chromosomes (n=3).

Instructions

1. Use one color of clay to make three different-sized "snakes" (one long, one medium, and one short)
2. Place the unduplicated chromosomes randomly in the center of your workspace.

Q1. You have just modeled 3 chromosomes. Which of the following terms can be applied to the set of chromosomes in your clay model? Check all that apply.

- ❏ duplicated chromosome
- ❏ homologous pair
- ❏ unduplicated chromosome

Q2. What stage of the cell cycle does your model currently represent? ___________________

3. Replicate your chromosomes by creating another, matching Play-Doh snake for each of the chromosomes already in your cell. Attach the two snakes by gently pinching them together somewhere along their length.

Q3. What process of the cell cycle is represented in this step? ___________________

Q4. How many pieces of DNA do you now have in your cell? ___________________

Q5. How many chromosomes are now in your cell? ___________________

Your model cell is now in prophase. During this stage of the cell cycle, DNA condenses, the nuclear envelope breaks down, and the spindle apparatus (mitotic spindle) begins to attach to the duplicated chromosomes.

4. Set aside half a page in your lab notebook.
5. Add a vertical line to split that section into two cells.
6. On the left, draw a diagram of a duplicated chromosome attached to the spindle.
7. Label that cell "Mitosis." The point is to emphasize how and where the spindle fiber(s) attach.
8. Set that sketch aside for use during meiosis modeling.
9. Line up your chromosomes in the center of the cell (single file, end-to-end).

Your cell is now in metaphase. The spindle fibers from opposite spindle poles "battle" until the duplicated chromosomes are all lined up.

10. Separate the sister chromatids of each of your chromosomes and begin to move them apart, toward their spindle poles.

You have just modeled anaphase. During this stage, the connections between the centromeres of a duplicated chromosome are broken. The motor inside the kinetochore (part of the centromere) starts walking the sister chromatids up the spindle fibers toward opposite spindle poles.

11. Move the unduplicated chromosomes to opposite spindle poles.

This is called telophase. When the unduplicated chromosomes (formerly sister chromatids) reach opposite spindle poles, they start to decondense, and a nuclear envelope forms around each cluster of chromosomes.

Q6. At what point in this process did duplicated chromosomes become unduplicated chromosomes? ____

Q7. How many chromosomes are present in each nucleus produced by this process? ______

Q8. What process would happen next (or at the same time as telophase)?

12. Compare the number of chromosomes in each of the daughter cells to the number of chromosomes in the original parent cell.
13. Repeat the process of modeling mitosis until you can do it without looking at your notes. You may want to photograph/sketch each stage.

Activity 2. Observing Mitosis in Plant Cells

Since the tip of a root is a region of rapid plant growth, the cells are dividing rapidly. This allows us to see cells of all the different stages of mitosis in one place. Think of each of the cells as a snapshot of one time in the cell cycle. Your job will be to identify cells in each of the stages of mitosis.

Figure 5.2. Microscope image of Allium root tip. Look for mitotic cells in the area indicated in the image.

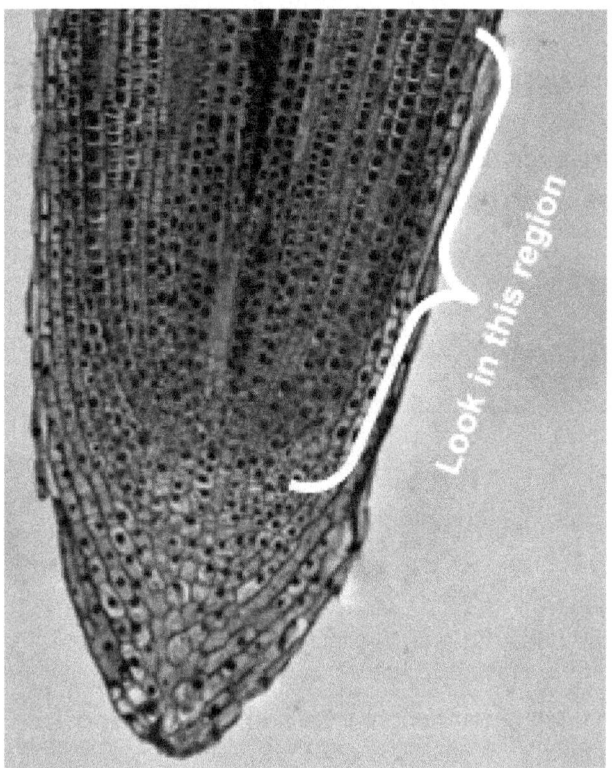

Instructions

1. Obtain a slide of Onion (*Allium*) Root Tips. This slide contains several slices of a root tip.
2. Focus on a region where there are cells with visible chromosomes. Refer to Figure 5.2 as you orient yourself to the slide.
3. Use the materials provided in class to identify cells at each stage of mitosis.
4. Find an example of an onion root tip cell undergoing cytokinesis.
5. Create scaled sketches of each stage of mitosis and cytokinesis in Allium in your lab notebook.

Tips for identifying mitotic stages in root tip cells

Interphase
- ❑ Cells containing a round or oval nucleus with an evenly stained nucleus.
- ❑ Most common cell stage in all images.
- ❑ Darker "balls" inside the nucleus are the nucleolus; there may be more than one.

Prophase
- ❑ Early prophase: the nucleus may still be oval, but has small or large splotches inside (condensing chromosomes).
- ❑ Late prophase: chromosomes appear to look more like noodles or snakes and are arranged in the shape of the nucleus (but no nuclear envelope present).

Metaphase
- ❑ Chromosomes form a line in the center of the cell.
- ❑ Noodle-like chromosome "legs" sticking out of the line.

Anaphase
- ❑ Chromosomes arranged in two clusters.
- ❑ Small space between the clusters, but no line (cell plate).

Telophase
- ❑ Chromosome clusters are further apart.
- ❑ DNA may be decondensing (losing noodle-like structure).
- ❑ A thin line (cell plate) exists in the cytoplasm between the chromosome clusters.

Activity 3. Which Stage of Mitosis is the Longest?

We have learned that most cells spend most of their time in interphase. The amount of time a cell spends in mitosis varies by species, but the percentage of time a cell spends in each stage of mitosis is fairly consistent.

In most tissues undergoing mitosis, cell division is asynchronous (i.e., it does not occur simultaneously in every cell), and it is often possible to see representatives of every stage in a single image.

As a result, if we take random "snapshots" of cells undergoing cell division and count the number of cells in each stage, we should be able to estimate which stage of mitosis is the longest (that would be the stage with the most representatives, right?).

Procedure A. Identifying and Counting Cell Stages

In this activity, you will use a database of images of cell division provided by the Allen Institute for Cell Science in Seattle, WA, to identify and count the number of animal cells in different stages of cell division.

QR Code 5.1. The Allen Institute for Cell Science

go.chemeketa.edu/Lab5Prelab

Using a little math, you will then determine the percentage of cells in each stage and create a graph to display and analyze your results. The more cells we count, the more accurate the results will be. This will be a group effort, where each student will be required to count cells in several images.

> Several resources are available on Canvas to help you with this exercise. Please check the week's pre-lab assignment on Canvas to access these links.

Instructions

1. Open the class mitosis spreadsheet using the link provided for this activity on Canvas.
   a. Column A contains hyperlinks to pictures of cells undergoing mitosis. Click on a link to see the image.
   b. You will claim 20 of these images by putting your initials in column B.
   c. In columns C through G, you will record the number of cells in each stage in your images.
   d. The total number of cells in each stage and the percent of cells in each stage are calculated by embedded formulas at the bottom of the spreadsheet (rows 929-932).
2. Claim the 20 images you plan to analyze by putting your initials in the indicated box.
3. Open the link provided for the first image to enter the cell viewer.
4. Expand the view and the tools by clicking on the icons at the top right of the page.
5. Identify the stage of and count all the cells with enough of the nucleus in the image to identify its stage. All cells with visible nuclei should be counted.
6. Enter the data directly into the database.
7. Wait until data from at least 100 images is added to the table to analyze the results.
8. Use the formulas at the bottom of the spreadsheet to determine the percent of cells in interphase and in each stage of mitosis.
9. Create a properly formatted graph of the data. Don't forget your figure legend.
10. Write a discussion section to support or reject the hypothesis you proposed in your pre-lab. Aim for approximately half a page. References not required.

5 | Applying What You've Learned

A selection of these questions will appear on your lab quiz. Lab quizzes are open book. Create digital answers to these questions in advance so that you can copy and paste the answers into the quiz and save time.

1. Describe how the process of cell division differs from the process of cytokinesis.
2. Describe one way that the structure of a chromosome in late prophase differs from the structure of a chromosome in G1 of interphase.
3. Several different drugs that affect the cytoskeleton are used to treat human diseases. Taxol and colchicine are examples that have opposite effects on microtubules. Use what you have learned to determine what stage of mitosis is affected by each drug and propose a mechanism for how the drug treats the health condition it is prescribed for.
   a. Taxol, a drug isolated from the Pacific Yew tree, is used as a treatment for breast cancer (defined as uncontrolled cell division). Taxol works by preventing the depolymerization of microtubules.
   b. Colchicine, a drug isolated from the autumn crocus (flower), is used as a treatment for a variety of medical conditions that cause inflammation (accumulation of white blood cells in locations like joints). Colchicine works by preventing the formation of microtubules.

6 | Gene Expression Pre-Lab

Instructions

- ❑ Read the lab manual and then follow the instructions to complete the assignment.
- ❑ You may need your textbook and other resources to complete this assignment.
- ❑ Pre-labs must be completed before the start of the lab.

Activity 1. Rock Pocket Mice

Use QR Code 6.1 to watch the video "The Making of the Fittest" and then answer the following questions.

QR Code 6.1. Natural Selection and the Rock Pocket Mouse

1. True or False? Add a T or F to the blank before each statement.

   _____ Mutations are caused by selective pressure in the environment.

   _____ The same mutation could be advantageous in one environment, but deleterious in a different environment.

   _____ The appearance of dark-colored volcanic rock caused the mutation for black fur to appear in the rock pocket mouse population

2. Which of the following contributes to selective pressure favoring dark fur in rock pocket mice? Select all that apply.
   - ❑ Predation
   - ❑ Rock Color
   - ❑ Mutation
   - ❑ Availability of food

3. Near the end of the film, Dr. Sean Carroll states that "while mutation is random, natural selection is not." What does he mean by this statement?

Activity 2. Protein Structure

4. Complete Table 6.1 by indicating the highest level of protein structure.

Table 6.1. Examples of High Levels of Protein Structures

Example	Level of protein structure (1°, 2°, 3° or 4°)
Beta-sheet	
Peptide bond	
Five proteins in a complex	
The active site of an enzyme	

Activity 3. Protein Structure

5. Complete Table 6.2 by adding X's or check marks to indicate the players involved in each step of a cell signaling pathway.

Table 6.2. Three Elements That Contribute to the Process of the Cell Signaling Pathway

	Reception	Signal Transduction	Response
May take place in the cytoplasm			
Requires proteins			
Involves a change in the function or appearance of a cell or organism			
May lead to a gene being turned on or off			
May involve a hormone			
May occur on the exterior of the plasma membrane			
Involves signaling molecules			

6 | Gene Expression

By the end of this lab, you should be able to:

- ❑ Differentiate between missense, nonsense, and silent mutations.
- ❑ Explain how a transmembrane protein can be used in cell signaling.
- ❑ Explain how the structure of a protein might change as a result of one or more point mutations.
- ❑ Predict the structure of a protein based on its amino acid sequence.
- ❑ Provide a basic explanation of how gene expression controls fur color in rock pocket mice.

The rock pocket mouse, *Chaetodipus intermedius*, is a small, nocturnal animal found in the deserts of the southwestern United States. Because most rock pocket mice have a sandy, light-colored coat, they are able to blend in with the light color of the desert rocks and sand on which they live. But populations of primarily dark-colored rock pocket mice have been found living in areas where the ground is covered with **basalt**, a dark rock formed by geologic lava flows thousands of years ago.

Scientists collected data from a population of primarily dark-colored mice living in basaltic terrain and from a nearby light-colored population. Analyzing DNA from these two populations revealed a mutation in the *MC1R* gene, which is involved in coat-color determination.

Wildtype MC1R

The MC1R protein is found in cells called melanocytes.
- ❑ **Melanocytes** produce pigments that affect the mouse's coat color.
- ❑ Fur color is a function of the presence of two pigments: **eumelanin** (dark) and **pheomelanin** (light).
- ❑ Wild-type fur is light because the melanocytes are low in eumelanin and high in pheomelanin.
- ❑ MC1R is one of several proteins that regulate the balance of eumelanin and pheomelanin.
- ❑ Mice with the wildtype *MC1R* gene have light fur.

The *MC1R* gene encodes a protein called melanocortin one receptor (MC1R).
- ❑ MC1R is a transmembrane protein with three regions.
- ❑ The **extracellular** region extends outside of the plasma membrane.
- ❑ The **intracellular** region extends into the cytoplasm of the cell.
- ❑ The **transmembrane** region is embedded within the plasma membrane.

MC1R and Signal Transduction

A simplified signal transduction pathway for MC1R is diagrammed in Figure 6.1. The numbers in the text correspond to the numbers in Figure 6.1.

- ❏ The MC1R protein contains an extracellular binding site for the hormone alpha-melano-cyte-stimulating (α-MSH).
- ❏ When α-MSH binds to the receptor (reception), MC1R changes its shape, causing the release of a G protein (signaling molecule; transduction), starting the transduction pathway.
- ❏ The G protein activates enzymes that increase the levels of cyclic adenosine monophosphate, or cAMP (another signaling molecule; transduction).
- ❏ cAMP activates a series of chemical reactions that turn on a gene (transduction).
- ❏ The gene is transcribed into an mRNA that moves to the cytosol (response)
- ❏ The mRNA is translated by a ribosome (response).
- ❏ The final product is an enzyme that catalyzes a chemical reaction, creating eumelanin (response).

Figure 6.1. A simplified signal transduction pathway for MC1R.

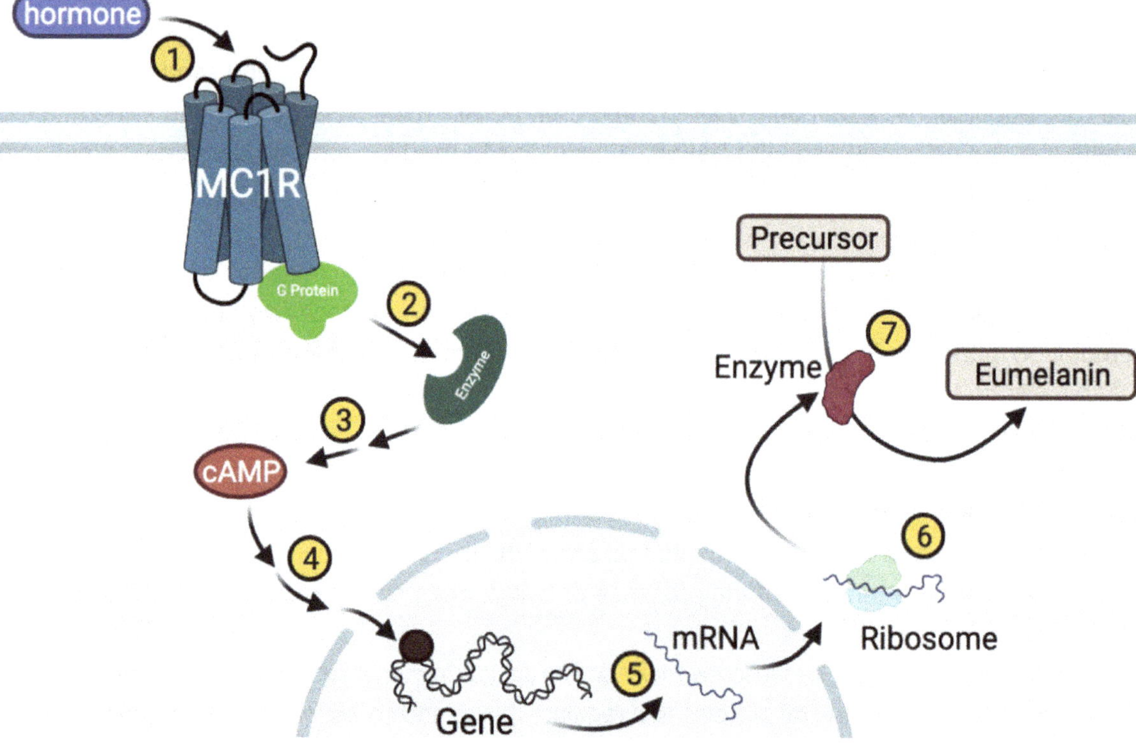

Q1. The phenotype of wild-type mice is light fur, but the signal transduction pathway described here turns on eumelanin (dark pigment) production. Why do wildtype mice have light colored fur?

Activity 1. Building Wild Type MC1R

In this activity, you will transcribe and translate a small portion of the *MC1R* gene. The actual gene contains 951 base pairs.

Procedure A. Transcribing and Translating mRNA

Instructions

2. Use the wild-type side of the "Gene Table" provided by your instructor to transcribe a section of mRNA.
3. Use the "Genetic Code" to translate that mRNA into a series of amino acids.

Q2. Based on what you know about the location of the transmembrane domain, what properties would you expect amino acids in that region of the protein to possess? Why?

4. Use the "Properties of Amino Acids Table" to determine the properties of each amino acid in both the wild-type and mutant proteins.
5. Color in each box of the two tables according to the color key below.
   a. Nonpolar (hydrophobic), neutrally charged amino acids: Green
   b. Polar (hydrophilic), neutrally charged amino acids: Blue
   c. Electrically charged, positive (basic) amino acids: Red
   d. Electrically charged, negative (acidic) amino acids: Yellow
6. Use a pipe cleaner and beads to model the structure of this section of the MC1R protein
7. Represent amino acids with pony beads. The color of the bead (green, blue, red, or yellow) represents the properties of the amino acids determined in the previous step.
8. Space the beads evenly along the pipe cleaner.
9. Use a piece of cardboard to create a model of the plasma membrane.
10. Fold the cardboard like a book.

The front cover is the top of the lipid bilayer, and the back cover is the bottom of the lipid bilayer. The spine is just holding them together. Maybe it represents the hydrophobic core of the membrane?

11. Label the extracellular and intracellular sides of the membrane
12. Inside the book, label the transmembrane region

13. Fold and insert the polypeptide into its three-dimensional shape in the membrane based on:
    a. Location of the domain in the cell (intercellular, extracellular, or transmembrane).
    b. The properties of the amino acids are defined by the different colored beads.
    c. The way you choose to represent these interactions is up to you, but be consistent
       in that method (and use the same method in the next exercise).
14. Use a hole punch to add holes to the membrane model as needed to insert the protein.

Q3. Examine the distribution of amino acids of different properties along the length of the
    polypeptide. What information can you get from the linear model at this point?

Q4. Do you notice any patterns? Clearly defined regions?

Q5. Are amino acids with the same properties mostly grouped together or spread out?

Q6. How does the distribution of amino acids of different properties affect protein
    structure?

Activity 2. Building Mutant MC1R

A mutation is any change in the DNA sequence of a gene. A mutation in a particular gene may (or may not) change the structure of the protein it encodes, potentially altering or preventing the protein's function.

Types of Mutations

- ❑ **Point Mutation:** Mutations that affect a single nucleotide.
- ❑ **Deletion:** The loss of one or more nucleotides from the DNA gene sequence. The deletion of nucleotides can result in frameshift mutations.
- ❑ **Insertion:** The addition of one or more nucleotides to the DNA gene sequence.
- ❑ **Substitution:** The replacement of one nucleotide of DNA with another.
- ❑ **Frameshift:** Caused by the insertion or deletion of a nucleotide. Changes the number and type of amino acids located after the mutation point.

Potential Results of a Mutation

- ❑ **Silent mutation:** Does not affect the amino acid sequence of the protein; therefore, there is no change in the resulting protein.
- ❑ **Missense mutation:** Causes an amino acid in the sequence to be changed to another amino acid. Changes the primary level of the protein structure and may impact protein conformation.
- ❑ **Nonsense mutation:** Causes the protein to be truncated (cut short) due to the incorporation of a "stop" signal into the DNA sequence. This results in translation being stopped before the amino acid sequence of the protein is completed.

Mutant MC1R

Dark color in the rock pocket mouse is the result of a mutation in the *MC1R* gene. When the protein produced by the mutant version of the gene is present, the production of eumelanin in melanocytes increases, resulting in the dark coat color.

Procedure A. Recording the Number of Amino Acids

Repeat the procedure used in Activity 1 to build a model of the mutant version of the *MC1R* gene.

Instructions

Complete Table 6.3 by indicating the number of the amino acid with a mutation and putting an "X" in the cell to indicate if the mutation was a silent, missense, or nonsense mutation.

Table 6.3. Determining the Number of Amino Acids with a Mutation

Amino Acid Number	Silent	Missense	Nonsense

Q7. Why is the mutation at amino acid location 211 not as significant as the other four mutations?

Q8. Why is it significant that the four missense mutations in MC1R are found in the extra-cellular and intracellular domains of the protein?

Activity 3. How Does MC1R Work?

The wildtype version of the *MC1R* gene produces light fur while the mutated form produces dark fur. When considering the signal transduction pathway that the MC1R protein regulates, you might find these phenotypes contrary to what you would expect to happen.

If *MC1R* controls the production of eumelanin, why aren't wild-type mice dark-furred and mutant mice light-furred?

Procedure A. Forming a Hypothesis

In this activity, you will work with your lab group to form a hypothesis about what occurs in the *MC1R* signal transduction pathway that results in the phenotypes observed in mice with the wild-type or mutant gene.

Instructions

1. Work with your group to develop a hypothesis about how the expression of the *MC1R* gene works in rock pocket mice. Refer to the information in this document and the models you produced as you work.
2. Use the icons in the slide deck provided by your instructor on Canvas to create diagrams that show how each version of the gene relates to the final appearance of the mouse.

This lab was adapted from HHMI-BioInteractive (2013), Molecular Genetics of Color Mutations in Rock Pocket Mice.

---

6 | Applying What You've Learned

A selection of these questions will appear on your lab quiz. Lab quizzes are open book. Create digital answers to these questions in advance so that you can copy and paste the answers into the quiz and save time.

1. How did the changes in the amino acid sequence in the mutant version of the *MC1R* gene affect the overall structure of the extracellular domain?
2. How did the changes in the amino acid sequence in the mutant version of the *MC1R* gene affect the overall structure of the intracellular domain?
3. How did the missense mutations in the *MC1R* gene affect the function of the MC1R protein?

The color of wild-type rock pocket mice is light because UV light cannot penetrate the mouse fur to initiate the *MC1R* signal transduction pathway.

4. Propose a hypothesis as to how mutations in the *MC1R* gene might cause mouse fur to be dark even when the cells are not exposed to UV light. *Hint: Use the MC1R slide deck to organize your thoughts.*

Name: _________________________   Lab Time: ________________   Due: ________________

<table>
<tr><td>**7**</td><td># Genetics Pre-Lab</td></tr>
</table>

Instructions

- ☐ Read the lab manual and then follow the instructions to complete the assignment.
- ☐ You may need your textbook and other resources to complete this assignment.
- ☐ Pre-labs must be completed before the start of the lab.
- ☐ Use QR Code 7.1 to watch the video "The Making of the Fittest" and then answer the following questions.

QR Code 7.1. Evolving Switches, Evolving Bodies.

go.chemeketa.edu/Lab7Prelab

Complete the following table based on what you learned about the stickleback fish in the video.

Table 7.1. Determining Fish Type Through Observing Pelvic Spines

Fish Type	Pelvic Spines (present/absent)
Marine (Oceanic)	
Freshwater (Bear Paw Lake)	

1. Which allele for pelvic spines is dominant? Why do you think this is the case?

A true-breeding oceanic stickleback and a true-breeding freshwater stickleback fish are crossed.

2. Based on the hypothesis you provided for Q2, what phenotype(s) would you expect to see in the offspring?

3. What ratio of stickleback fish with spines to the stickleback fish without spines would you expect to observe?

4. If you cross two F_1 sticklebacks, what would the ratio of stickleback fish with spines to the stickleback fish without spines would you expect to observe?

5. If the F_2 generation included 40 offspring, how many would you expect to have spines, and how many would you expect to lack spines?

Read the information provided on the Chi-square test on Canvas and answer the following questions.

6. What is tested by the Chi-square test?

7. The formula for calculating the Chi-square is $\chi^2 = \Sigma\ (o-e)2\ /e$. What do each of the symbols in this formula represent?

    $\Sigma =$ $o =$ $e =$

8. What in a genetic cross would be considered the expected data?

9. What is represented by the p-value in a Chi-square test?

10. How will you use the Chi-square test to determine if your data supports your hypothesis in the genetics lab?

7 | Genetics

The study of inheritance, known as genetics, focuses on how traits are inherited by offspring. Geneticists study one or a few genes to understand how different alleles of a gene interact, and sometimes how those alleles interact with the alleles of other genes. In this lab, we will study genetics in stickleback fish and explore how genetic counselors trace inheritance over many generations in humans using pedigrees.

By the end of this lab, you should be able to:

- ❏ Use the Chi-square test to analyze data from a genetic cross.
- ❏ Use Mendelian principles to predict the outcomes of genetic crosses following one or two independently assorting genes.

Activity 1. Genetics of Pelvic Spine Inheritance in Stickleback Fish

Geneticists breed organisms with different phenotypes to answer questions about how phenotypes are inherited. Is one phenotype dominant and one recessive? Is each phenotype controlled by a different version (allele) of a single gene, or are many interacting genes involved?

In this activity, you will answer these questions by analyzing the outcome of breeding a stickleback with pelvic spines and one without pelvic spines.

Figure 7.1. Threespine stickleback. In one allele (shown), the pelvic spine is present. In the other allele pelvic spines are absent. Ruler units are cm.

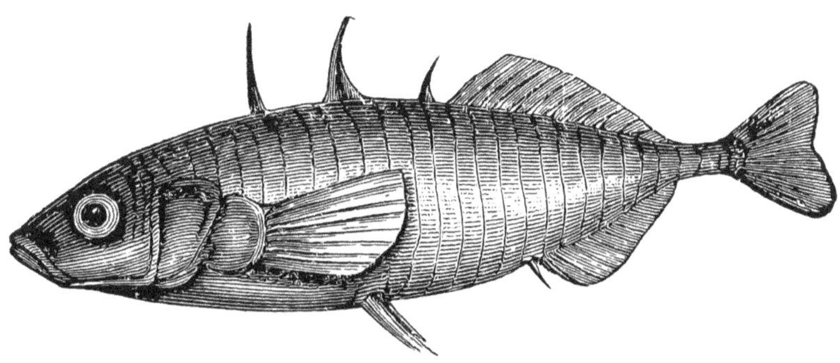

Pelvic Spine Phenotypes

The fish shown in Figure 7.1 is a marine three-spined stickleback. Stickleback fish have a pair of pelvic spines (only one is visible in the photo), which serve as a defense from large predatory fish. In some freshwater populations, stickleback fish lack pelvic spines.

Answer these questions to get a set of fish cards:

Q1. What are the two phenotypes for spines in stickleback fish?

Q2. How would you determine which of these alleles is dominant?

Q3. Which allele do you think is most likely to be dominant and why?

Activity 2. Gathering Data

In this activity, you will examine the offspring (F_1 and F_2) of a genetic cross between an ocean stickleback and a freshwater stickleback.

Instructions
1. Sort the F_1 set of cards into two separate piles: fish with pelvic spines and fish without pelvic spines
2. Repeat the same procedure with the F_2 set of cards.
3. Count and record the total number of fish with each phenotype in Table 7.2.

Table 7.2. **Results of a Cross Between Marine and Bear Paw Lake Stickleback**

Generation	Fish With Pelvic Spines	Fish Without Pelvic Spines
P	1	1
F_1		
F_2		

Q4. What is the ratio of fish with pelvic spines to fish without pelvic spines in the F_1 generation?

Q5. What is the ratio of fish with pelvic spines to fish without pelvic spines in the F_2 generation?

Q6. According to these results, which phenotype is dominant and which is recessive? Explain using evidence.

Procedure B. Analyzing Additional Experimental Data

In your crosses, you only looked at a small number of offspring. Researchers usually perform many more crosses and count many more offspring. Tables 7.3 and 7.4 contain the results of several other crosses. Freshwater stickleback fish were collected from several Alaskan lakes (Bear Paw Lake, Boot Lake, and Whale Lake).

Instructions

1. Add your data from the F_1 fish cards from Table 7.2 to the "Your Cross" row of Table 7.3. Complete Table 7.3 by calculating the total F_1 offspring with and without pelvic spines.
2. If you look at the total number of F_1 fish, what is the ratio of fish with pelvic spines to fish without pelvic spines?
3. Add your data (total and phenotypic ratio) to the appropriate cells of Table 7.4.
4. Calculate the ratio of fish with pelvic spines to fish without pelvic spines (for example, 3:1 or 2.8:1) for each family and for the total population.
5. Round your numbers to the first decimal place.
6. Record these calculations in column D.

Parental Cross	F_1 Generation Total Offspring	With Pelvic Spines	Without Pelvic Spines
Your Cross			
Marine x Bear Paw Lake	50	50	0
Marine x Boot Lake	33	33	0
Marine x Whale Lake	42	42	0
Total			
Phenotypic Ratio			

Source: 2004 Proc. Natl. Acad. Sciences 101: 6050-6055.

Q7. Did Dr. Cresko's data for the F_1 generation agree with your F_1 data? Was this expected?

Q8. Did the addition of your data for the F_1 generation to Dr. Cresko's data change the ratio of fish with and without pelvic spines in the F_1 generation?

Q9. Look at the phenotype ratio for all of the crosses combined (bottom of column D in Table 7.3). What does that ratio tell us about the inheritance of the pelvic spine allele?

Q10. Notice that the phenotype ratios vary from family to family. Explain why every family
    does not show the same ratio.

Table 7.4. Crosses Between F_1 Generation Stickleback from the Marine x Bear Paw Lake Parental Cross: The F_2 Generations

F_2 Generation	(A) Total Offspring	(B) With Pelvic Spines	(C) Without Pelvic Spines	(D) Ratio of Pelvic Spines to No Pelvic spines
Your cross				
Family 1	98	71	27	
Family 2	79	62	17	
Family 3	62	49	13	
Family 4	34	28	6	
Family 5	29	24	5	
Family 6	23	17	6	
Family 7	21	17	4	
Family 8	19	18	1	
Family 9	15	11	4	
Family 10	12	10	2	
Family 11	12	10	2	
Family 12	4	3	1	
Total				

Procedure C. Chi-Square Statistics

So far, you have evaluated whether the ratio of fish with spines to fish with no spines suggests
that the inheritance of pelvic spines follows Mendel's Law (one gene with complete dominance).
Likely, your results did not exactly match your predicted ratios. Were they close? How close is
close enough?

In this activity, you will analyze your data to determine if the difference between expected results
and actual results is scientifically significant. Review the material provided about performing the
Chi-Square test provided in this week's lab activity on Canvas before starting.

1. Determine the expected number of offspring with and without spines if this gene follows Mendel's laws (expected number)
2. Determine the observed number of offspring with and without spines (Table 7.4 totals).
3. Complete Table 7.5 to determine the Chi-square value.

Table 7.5. Chi-Square Analysis of Stickleback Data (monohybrid)

	Pelvic Spines	No Pelvic Spines
Observed Value (*o*)		
Expected Value (*e*)		
(o-e)		
$(o-e)^2$		
$(o-e)^2/ e$		
Σ		

Q11. What are the degrees of freedom (df) for this analysis?

Q12. What is the p-value at 0.05 using this degree of freedom? Use the Critical Values Table on the Chi-Square page of this week's module.

Q13. Does the data in Table 7.5 support the hypothesis that the pelvic spine gene follows Mendel's laws?

Activity 3. Dihybrid Crosses

Stickleback fish can also exhibit differences in body armor. Marine sticklebacks tend to have body armor, while some freshwater sticklebacks do not. Data suggests that the "no armor" allele developed several times, but one of those cases is inherited with a 3:1 ratio, suggesting that the trait is controlled by a single gene that follows the laws of Mendelian inheritance. The allele that expresses armor is dominant.

To determine if the pelvic spine gene and the armor gene assort independently, a true-breeding fish with pelvic spines and armor is crossed to a true-breeding fish without pelvic spines or armor. Offspring from this cross (the F_1 generation) were crossed to produce an F_2 generation containing 871 offspring. Of those offspring, 490 had pelvic spines and armor, 169 had pelvic spines and no armor, 157 had no pelvic spines and armor, and 55 had no pelvic spines and no armor.

Q14. What would the finding that these genes are independently assorting indicate about their location in the stickleback genome?

Q15. Choose letters to represent these two genes and create a key to the alleles. What is the genotype of the P generation used in this cross?

Q16. What are the genotypes and phenotypes of the F_1 generation?

Q17. What gametes do the F_1 generation produce?

Q18. What will the results of crossing two F_1 fish be if the genes assort independently?

Instructions

1. Complete the Punnett Square (Table 7.6).
2. Place possible gametes on the sides and tops of the table.
3. Determine all possible combinations of offspring produced by this cross.
4. Below the genotype of each offspring, indicate its phenotype.

5. Count the number of squares represented by each phenotype combination. Record your numbers in Table 7.7. These numbers form the expected phenotypic ratio (put colons in between them).
6. Remembering that the total offspring produced in the cross is 871, use the ratio to calculate the expected number of offspring of each type (Table 7.8).
7. Add the observed offspring (listed in the introduction to this exercise) to Table 7.8
8. Complete Table 7.8 to calculate the Chi-Square value for this cross.

Table 7.6. Punnett Square to Determine Possible Offspring in the F$_2$ Generation

Gametes				

Table 7.7. Expected Phenotype Distribution in F$_2$ Generation

Phenotype	Number of cells with phenotype (in Punnett Square)
Pelvic spines + Armor	
Pelvic spines + No armor	
No pelvic spines + Armor	
No pelvic spines + No armor	

	Possible Phenotype Combinations			
Observed Value (o)				
Expected Value (e)				
(o-e)				
(o-e)2				
(o-e)2/ e				
Σ				

Q19. Based on the results of the Chi-Square analysis, does the data agree or disagree with the hypothesis that the pelvic spine and armor genes are independently assorting genes that follow Mendel's laws? Use the data and your Chi-Square analysis to provide a scientific explanation for your answer.

Activity 4. Pedigree Analysis

A pedigree is a visual representation of how a particular trait is inherited from one generation to the next. Using pedigrees is like solving logic problems. You have to think your way through. It's often helpful to start at the bottom and work your way up. Figure 7.2 provides a key to interpreting pedigrees.

Figure 7.2. A key to decoding pedigrees.

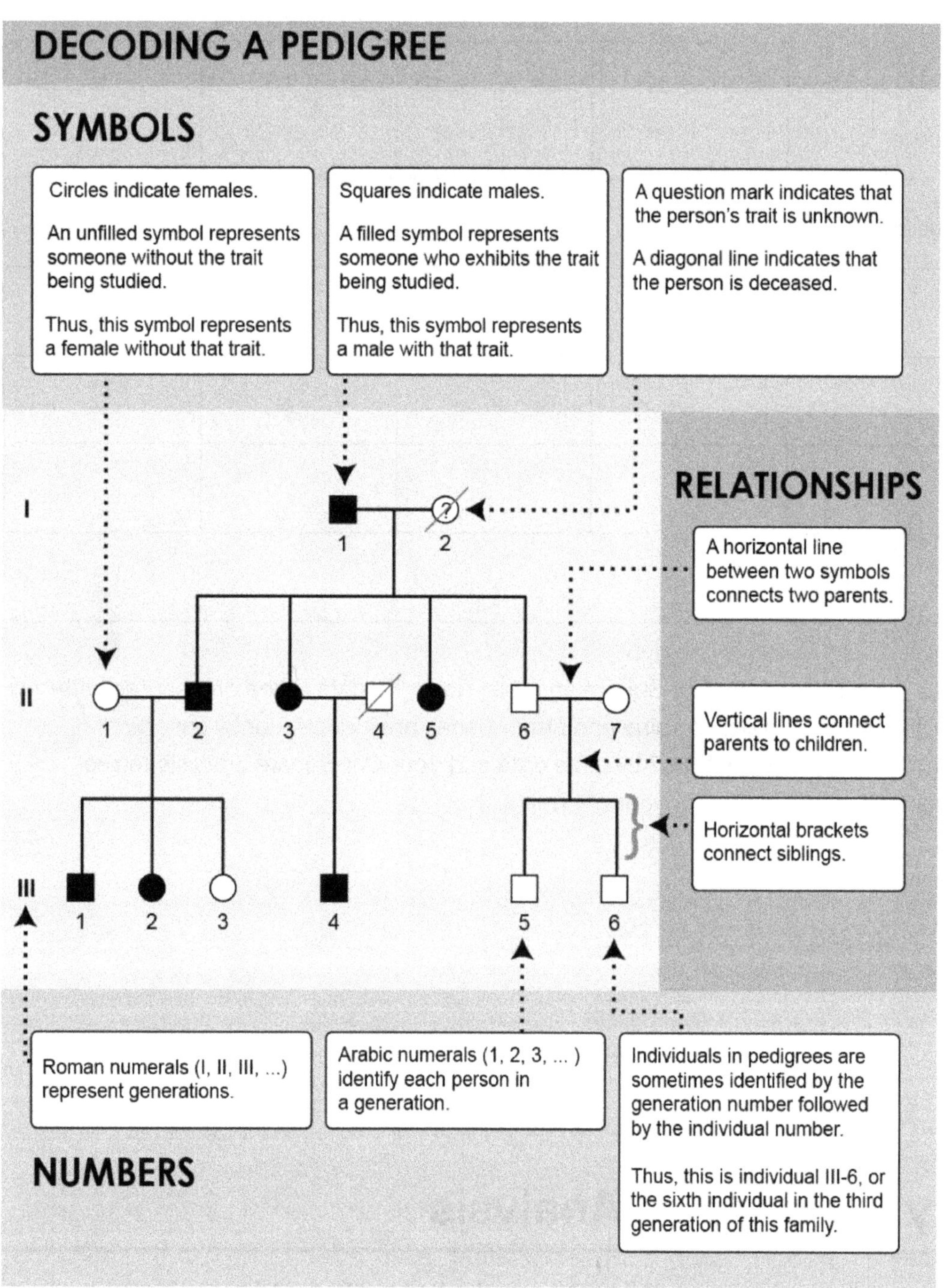

DECODING A PEDIGREE

SYMBOLS

Circles indicate females.
An unfilled symbol represents someone without the trait being studied.
Thus, this symbol represents a female without that trait.

Squares indicate males.
A filled symbol represents someone who exhibits the trait being studied.
Thus, this symbol represents a male with that trait.

A question mark indicates that the person's trait is unknown.
A diagonal line indicates that the person is deceased.

RELATIONSHIPS

A horizontal line between two symbols connects two parents.

Vertical lines connect parents to children.

Horizontal brackets connect siblings.

Roman numerals (I, II, III, ...) represent generations.

Arabic numerals (1, 2, 3, ...) identify each person in a generation.

Individuals in pedigrees are sometimes identified by the generation number followed by the individual number.
Thus, this is individual III-6, or the sixth individual in the third generation of this family.

NUMBERS

Procedure A. Using Pedigrees to Determine Mode of Inheritance*

The film Got Lactase? The Co-evolution of Genes and Culture traces the evolution of lactose tolerance and describes how researchers analyzed the pedigrees of several Finnish families to identify the changes in the DNA mutations responsible for this trait. Use QR Code 7.2 to watch the film.

QR Code 7.2. Got Lactase? The Co-evolution of Genes and Culture

https://www.youtube.com/watch?v=MA9bol1qTuk

In this activity, you will analyze some of those pedigrees to determine how the lactose-tolerant trait is inherited. You will then analyze DNA sequences to identify mutations associated with the trait.

**Adapted from a figure in Enattah, N. S., et al. 2002 Nature Genetics 30: 233-237*

Instructions

Use the pedigrees in Figure 7.3 to answer the following questions.

Figure 7.3. The lactose tolerance chart for Family A.

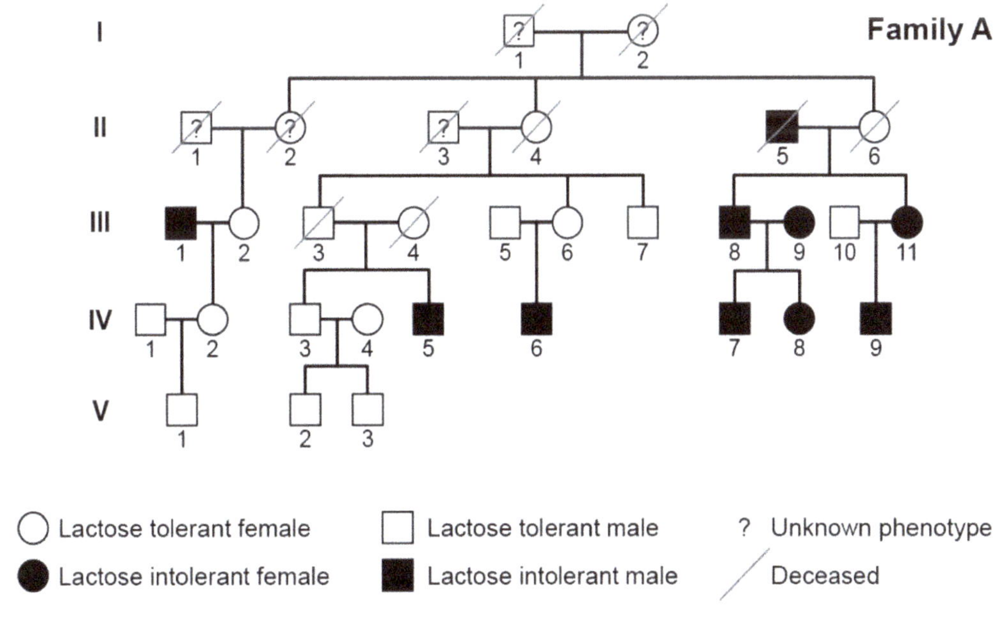

Q20. Is lactose tolerance a dominant or recessive trait? Provide a scientific explanation supporting your answer using at least two individuals from Family A to support your claim.

Q21. Individuals 5 and 6 in generation III of Family A are both lactose-tolerant, yet their son is lactose-intolerant. Does this information agree with the claim you made in the previous question? Use the phenotype of at least two other individuals in Family A to support your answer.

Procedure B. Finding the Mutation

Different alleles result from different DNA sequences. Different DNA sequences result from random mutations.

Tables 7.9 and 7.10 show DNA sequences from two different regions of DNA on chromosome 2 that show variation between individuals.

- ❑ Because each individual has two copies of chromosome 2 (one from each parent), each row includes two DNA sequences (one from each of their chromosomes).
- ❑ Individuals are identified by family letter, generation number, and individual number. For example, Individual BIV-4 is Individual 4, Generation IV in Family B.

Instructions

1. Use the pedigrees (Figures 7.3 and 7.4) to determine the phenotype of each individual and complete the Phenotype column in both Table 7.9.
2. The same individuals are used in Table 7.10. Copy the phenotype column from Table 7.9 to the phenotype column in Table 7.10.
3. Circle or highlight all the nucleotides that differ between sequences in both tables.

Figure 7.4. Lactose tolerance charts for Families B, C, and D.

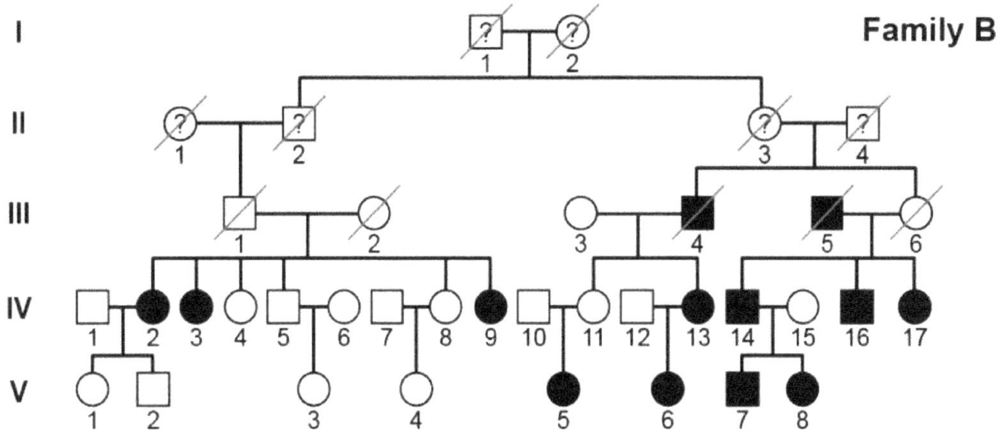
Family B

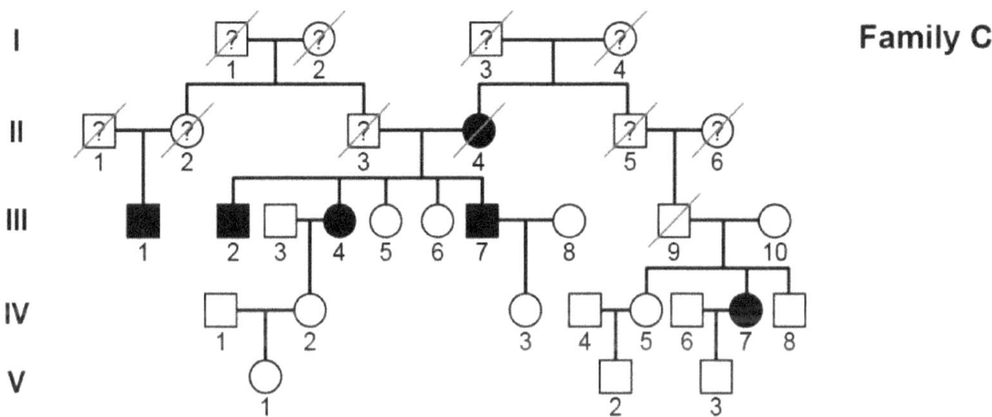
Family C

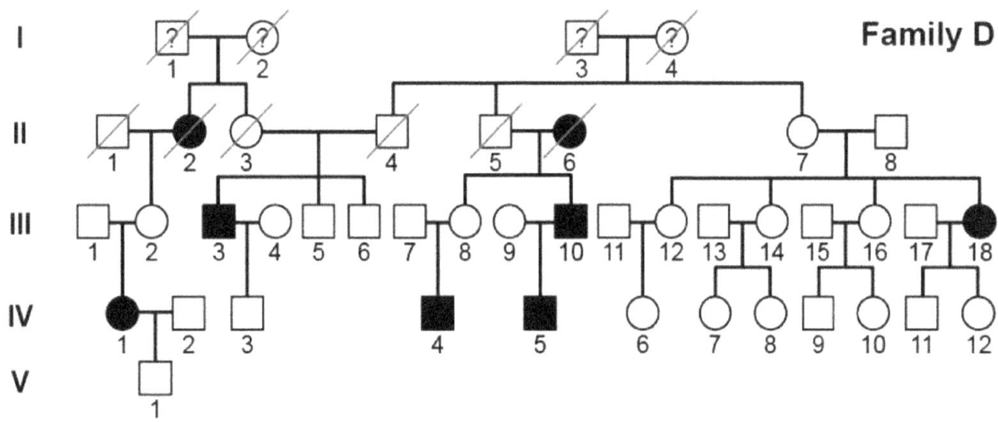
Family D

Lactose tolerant female
Lactose intolerant female
Lactose tolerant male
Lactose intolerant male
? Unknown phenotype
Deceased

Table 7.9. DNA Sequence Variation #1 from Lactose Tolerant and Lactose Intolerant Individuals

Individual	Phenotype	Sequence 1*
A IV-3		*Copy 1, Chromosome 2:* TAAGATAATGTAGTCCCTGG
		Copy 2, Chromosome 2: TAAGATAATGTAGTCCCTGG
B IV-4		*Copy 1, Chromosome 2:* TAAGATAATGTAGTCCCTGG
		Copy 2, Chromosome 2: TAAGATAATGTAGTCCCTGG
B IV-8		*Copy 1, Chromosome 2:* TAAGATAATGTAGTCCCTGG
		Copy 2, Chromosome 2: TAAGATAATGTAGCCCCTGG
B IV-9		*Copy 1, Chromosome 2:* TAAGATAATGTAGCCCCTGG
		Copy 2, Chromosome 2: TAAGATAATGTAGCCCCTGG
C IV-3		*Copy 1, Chromosome 2:* TAAGATAATGTAGTCCCTGG
		Copy 2, Chromosome 2: TAAGATAATGTAGCCCCTGG
D IV-4		*Copy 1, Chromosome 2:* TAAGATAATGTAGCCCCTGG
		Copy 2, Chromosome 2: TAAGATAATGTAGCCCCTGG

Sequence variation #1 corresponds to nucleotides 13923-13902 upstream from the start of the lactase gene.

Table 7.10. DNA Sequence Variation #2 from Lactose Tolerant and Lactose Intolerant Individuals

Individual	Phenotype	Sequence 2*
A IV-3		*Copy 1, Chromosome 2:* ATAAAGGACACTCTTGACAA
		Copy 2, Chromosome 2: ATAAAGGACACTCTTGACAA
B IV-4		*Copy 1, Chromosome 2:* ATAAAGGACACTCTTGACAA
		Copy 2, Chromosome 2: ATAAAGGACACTCTTGACAA
B IV-8		*Copy 1, Chromosome 2:* ATAAAGGACACTCTTGACAA
		Copy 2, Chromosome 2: ATAAAGGACGCTCTTGACAA
B IV-9		*Copy 1, Chromosome 2:* ATAAAGGACGCTCTTGACAA
		Copy 2, Chromosome 2: ATAAAGGACGCTCTTGACAA
C IV-3		*Copy 1, Chromosome 2:* ATAAAGGACACTCTTGACAA
		Copy 2, Chromosome 2: ATAAAGGACACTCTTGACAA
D IV-4		*Copy 1, Chromosome 2:* ATAAAGGACACTCTTGACAA
		Copy 2, Chromosome 2: ATAAAGGACACTCTTGACAA

Sequence variation #2 corresponds to nucleotides 30192-30173 upstream of the start of the lactase gene.

Q22. What variations did you find in sequence variation #1?

Q23. What variations did you find in sequence variation #2?

Q24. Why did we look at two different sequence variations in these individuals?

Q25. Based on the pedigrees and sequence variation data, which sequence variation is associated with lactose tolerance? How did you make this determination?

Q26. Does the other sequence variation relate to lactose tolerance? Explain your answer.

7 | Applying What You've Learned

A selection of these questions will appear on your lab quiz. Lab quizzes are open book. Create digital answers to these questions in advance so that you can copy and paste the answers into the quiz and save time.

1. Do the results of Activity 1 (following the inheritance of pelvic spines) support the hypothesis that a single gene controls this phenotype? Explain using the data from your cross.
2. Do the results of Activity 2 (following the inheritance of pelvic spines and armor) support the hypothesis that these phenotypes are controlled by two different genes?
3. Does an individual with the genetic variation associated with lactose tolerance have to have two copies of that allele to be lactase persistent? What does that tell you about the lactose tolerance allele?
4. Recall from the video we watched on the evolution of lactose-tolerance that the phenotype of lactose-tolerance evolved more than once. Does the data in Tables 7.9 and 7.10 provide evidence for this claim?
5. Examine the pedigrees for Families B, C, and D. Do these families support your hypothesis about the inheritance of lactose tolerance?

L is the symbol for the lactose tolerance allele, and l is the symbol for the lactose intolerant.

6. What are the possible genotypes of:
7. An individual who is lactose tolerant:
8. An individual who is lactose intolerant:
9. Determine the genotype of Individual 3, generation IV in Family C? How did you make this determination?
10. Determine the genotype of the father of Individual 4, generation IV in Family D? How did you make this determination?
11. Individuals 1 and 2, generation IV in Family D are about to have a second child. What is the probability that their second child will be lactose intolerant? How did you make this determination?

8 | CRISPR Out of the Blue Pt 1 Pre-Lab

Instructions

- ❑ Read the lab manual and then follow the instructions to complete the assignment.
- ❑ You may need your textbook and other resources to complete this assignment.
- ❑ Pre-labs must be completed before the start of the lab.

Use QR 8.1 to watch an introductory video, "Genome Editing with CRISPR-CAS9," and review the student guide provided on the Canvas page for this assignment to answer the following questions.

QR Code 8.1. Genome Editing with CRISPR-CAS9.

youtube.com/watch?v=2pp17E4E-O8&feature=youtu.be

You may find this difficult. That is okay! Remember that we will work through this in the lab. Just do your best to think things through right now. That will make sure you are better prepared for the lab.

1. What is the job of the *LACZ* gene in bacteria?

2. What happens when bacteria with an active *LACZ* gene are fed X-gal instead of lactose?

Table 8.1. Characteristics of Bacterial Plates

Starter Plate	Additives	Bacterial Colony Color	CAS9
IX	IPTG, X-gal	Blue	+
IX/ARA	IPTG, X-gal, arabinose	Blue	+

3. Why does IPTG have to be present in both plates?

4. Why are the bacteria in both starter plates colored blue?

5. What color would the bacterial colonies on these plates be if their *LACZ* gene was NOT functional. Why?

6. How do the IX and IX/ARA starter plates differ?

7. What happens if the bacteria on the IX plate experience a double-stranded DNA break? How does this differ in the IX/ARA plate?

8. The bacteria we will grow on both starter plates have the gene for the CAS9 enzyme.
    Does this mean they can perform CRISPR without any additional additives? Explain

The additives used in this experiment are listed below. Complete Table 8.2 by indicating what role
each additive plays in the experiment we will perform.

Table 8.2. The Roles of Additives

Additive	Role in Experiment
IPTG	
X-gal	
ARA	
pLZDonor	
pLZDonor Guide	

9. What is the difference between the two plasmids we will use in this experiment?

10. Can the CAS9 enzyme cut bacterial DNA if the pLZDonor plasmid is present? Why or
    why not?

11. Can the CAS9 enzyme cut bacterial DNA if the pLZDonor Guide plasmid is present? Why or why not?

12. Considering all the components of this experiment, explain what must happen for bacteria to successfully undergo gene editing? How will you know if the bacteria have been edited?

8 | CRISPR Out of the Blue: Pt 1

This 3-part lab is designed to give you an introduction to modern molecular biology techniques. Molecular biology is used to study the structure and function of DNA and proteins in the cell. More recently, molecular biology techniques that allow DNA to be altered or edited have assisted scientists with stem cell research and the development of a variety of genetically modified organisms that support our society.

By the end of this lab, you should be able to:
- ❑ Accurately operate a micropipette.
- ❑ Demonstrate how to inoculate a prepared plate with bacteria.
- ❑ Describe the process of gene editing at a basic level.
- ❑ Explain the role of color indicators and antibiotic resistance in molecular biology.

Activity 1. Simulate the Molecular Mechanism of CAS9 DNA Cleavage

Use the paper model to walk through the steps of CRISPR-CAS9 DNA cleavage using a sequence from the bacterial gene *LACZ* which encodes b-galactosidase. The *LACZ* gene is part of the lac operon, a collection of genes that allows bacteria to use lactose, a milk sugar, as a food source. The DNA and sgRNA sequences in the paper model match those used in Activity 2,

Instructions

1. Cut out the sgRNAs and DNA strips. You may leave the CAS9 protein on its page.
2. Use the steps in Figure 2 as a guide to model the CRISPR-CAS9 mechanism:
    A. CAS9 binds an sgRNA: Place sgRNA 1 onto the CAS9 illustration and align it with the dotted lines.
    b. The CAS9-sgRNA complex binds to a PAM site. Place DNA strip 1 on the stripe across the CAS9 model. Slide the DNA strip until the PAM box on the CAS9 protein matches a PAM (5I-NGG) sequence on the DNA.
    c. The guiding region of the sgRNA binds to the target DNA sequence. Check whether the DNA sequence is complementary to the sgRNA sequence (U pairs with A, C pairs with G). If they are complementary, continue the process. Otherwise, repeat steps 2.b and 2.c with a new PAM site.

D. CAS9 makes a double-stranded break in the DNA: The scissors icons indicate where CAS9 cuts the DNA strands. Use a pencil to draw a vertical line across both strands at this position.

3. Verify that you have chosen the correct cut site and then use a pair of scissors to cut DNA strip 1 at that site. Keep the pieces of DNA strip 1 for use in Part 3.

Q1. How many nucleotides long is the guiding region of the sgRNA?

Q2. Does the sgRNA bind to the PAM?

Q3. Where does CAS9 cut the target DNA relative to the protospacer sequence?

8

Activity 2. Design the Guiding Region of an sgRNA

CRISPR technology is powerful in part because the target DNA sequence is controlled by a customizable sgRNA. In this activity, you will customize the guiding region of the sgRNA to cut a target site on the *LACZ* gene. DNA strip 2 represents a DNA sequence from the *LACZ* gene where you wish to make an edit.

Instructions

1. Use sgRNA 2, DNA strip 2, and the steps you followed in Part 1 to determine the sgRNA guiding region sequence required to direct CAS9 to cut DNA strip 2 at the red dashed line.
2. Write the nucleotide letters (A, U, C, G) of this sequence into the spaces on sgRNA 2. Use the steps of CAS9 DNA cleavage to confirm that the sequence you wrote on sgRNA 2 is correct.
3. Indicate your final sequence in Table 8.3.

sgRNA Guiding Region Sequence

Q4. How does the requirement of a PAM sequence affect the flexibility of CRISPR-CAS9 gene editing?

Q5. How would you identify a target DNA cleavage site for CRISPR-CAS9 and design an sgRNA?

Activity 3. Design Donor Template DNA for DNA Repair

CRISPR-CAS9 can find a specific sequence in a genome billions of base pairs long and then cut at a precise location within that sequence. How do scientists and researchers use the specificity of CRISPR-CAS9 to direct targeted gene editing? When chromosomal DNA in a bacterial cell is cut, the cell will die unless it's able to repair the cut. Bacteria have evolved processes to repair double-strand DNA breaks that would otherwise lead to cell death.

In this activity, we will explore Homology directed repair (HDR), where enzymes patch the break using a specific donor template DNA. The templates you create will contain a specific DNA sequence that will be inserted into the host DNA, flanked on each side by 15-bp sequences (homology arms) that are complementary to the sequence on each side of the cut made by CRISPR-CAS9.

Procedure A.

Instructions

1. Retrieve the two pieces of DNA strip 1. If you have not already used scissors to cut the strip at the cut site, do so now.

2. Cut out the donor template DNA strip and two blank DNA strips.

3. The shaded region of the donor template represents the sequence of DNA that we want to add to the donor DNA.

4. In the empty boxes on either side of the shaded region and on both strands, write the 15 bp sequences that match the nucleotide sequences on either side of the cut site of DNA strip 1.

5. These 15 bp sequences are your homology arms.

6. You now have a complete donor template DNA.

7. Use scissors to cut the excess ends of the donor template strip (remember that each homology arm is only 15 bp long).

8. Place the pieces of DNA strip 1 directly on top of the donor template strip nucleotide sequences so that the homology arms are aligned and only the insertion sequence is visible. Tape the pieces together.

You now have an edited piece of DNA.

9. Use the blank DNA strip along with the rest of the paper model pieces to design donor template sequences with 15 bp homology arms that will induce each of the following changes to DNA strip 1 using sgRNA 1:
   a. Cause a frameshift
   b. Insert an EcoRI restriction site (GAATC)

10. Add your final sequences to Table 8.4. Include the insertion sequences and the homology arm sequences. Underline the homology arm regions.

Table 8.4. Recording Final Sequences

Desired Change	Donor Template DNA Sequence
Cause a frameshift	
Insert an EcoRI restriction site	

Q6. How would you use CRISPR-CAS9 and HDR to treat a genetic condition caused by a single point mutation?

Activity 4. Perform Gene Editing

Instructions

1. Record your prediction for the outcome of your experiment in Table 8.5.
2. Draw the table into your laboratory notebook or include a scan of this table in your laboratory notebook.
3. Follow steps 1-17 in the gene editing protocol provided by your instructor.

Table 8.5. Bacterial Plate Cultures

Plate	Additives					Growth Expected? (YES/NO)	Color of Colonies (if growth occurs)
	IPTG	X-Gal	ARA	pD	pDg		
A	✓	✓		✓			
B	✓	✓			✓		
C	✓	✓	✓	✓			
D	✓	✓	✓		✓		

Name: _______________________________ Lab Time: ________________ Due: ___________________

8 | Applying What You've Learned

A selection of these questions will appear on your lab quiz. Lab quizzes are open book. Type up answers to these questions in advance so that you can copy and paste the answers into the quiz.

1. Why is colony color an indicator of *LACZ* gene expression?
2. What happens to a bacterium if a double-stranded DNA break is not repaired?
3. What is the purpose of using both the IX and IX/ARA starter plates in this experiment?
4. Describe at least two ways that the CRISPR-CAS9 system might be used for gene therapy (in general rather than for a specific disease).
5. Describe at least two different risks associated with using the CRISPR-CAS9 system for gene therapy.

9 | CRISPR Out of the Blue Pt 2 Pre-Lab

Instructions

- ❏ Read the lab manual and then follow the instructions to complete the assignment.
- ❏ You may need your textbook and other resources to complete this assignment.
- ❏ Pre-labs must be completed before the start of the lab.

Examine the results of your group's CRISPR experiment (photos provided by instructor) and complete the following questions.

Activity 1. Making a Prediction

Predict the results of the gene editing activity (CRISPR Part 1) by completing Table 9.1. Complete each empty cell by indicating whether bacteria should grow, and if they did, what color the colonies should be.

Table 9.1. Bacterial Growth and Colony Colors

	pD + Donor DNA	pDG + Donor DNA and Guide RNA
IX Plate		
IX-ARA Plate		

1. Which plates show evidence of the *LACZ* gene being cut by CAS9? Use your data to support your answer.

2. Do any of the plates show evidence of *LACZ* DNA being cut but not repaired? Use your data to support your answer.

3. Do your answers agree with the predictions you made during the CRISPR Part 1 Lab? Why or why not?

Activity 2. Polymerase Chain Reaction

Use QR Code 9.1 to watch an introduction to PCR and review the PDF provided by your instructor on Canvas to prepare for this assignment.

QR Code 9.1. PCR (Polymerase Chain Reaction)
https://www.youtube.com/watch?v=a5jmdh9AnS4

4. What is the goal of Polymerase Chain Reaction (PCR)?

5. Why is it important to use an enzyme like Taq Polymerase to perform PCR?

6. What is the role of a primer in PCR?

7. What is an amplicon?

Considering what you have learned about the experiment we will perform in this week's lab:

8. Why is it unlikely that both a 1,100 bp and a 650 bp amplicon would be produced in one of your samples?

9. What would you conclude about the *LACZ* gene if PCR produced only a 350 bp amplicon?

10. What would you conclude about the *LACZ* gene if PCR did not result in a 350 bp amplicon?

9 | CRISPR Out of the Blue: Pt 2

In the first part of this lab, we used the CRISPR system to disrupt the *LACZ* gene in *E. coli*. Bacteria with an active *LACZ* gene, when fed X-gal, produce blue colonies. In our experimental system, bacteria that were successfully edited by CRISPR are unable to produce a functional *LACZ* protein and, therefore, produce white colonies. In this lab, we will double check our work by using polymerase chain reaction (PCR) to amplify fragments of DNA from modified and unmodified bacteria using three sets of primers. One set of primers will amplify a region of the unmodified *LACZ* gene. The second set of primers will amplify a region of the modified gene. The final primer acts as a control as it amplifies a fragment of DNA that is not involved in lactose metabolism.

By the end of this lab, you should be able to:
- ❏ Demonstrate the proper usage of a micropipette.
- ❏ Explain how a single copy of DNA can be amplified by polymerase chain reaction (PCR).
- ❏ Predict the results of PCR using primers that produce a DNA fragment of a specific size.

Activity 1. Analyzing CRISPR Results

In the first part of this laboratory, you performed CRISPR to disrupt the expression of the *LACZ* gene in *E. coli*.

Q1. Did your experiment work?

Procedure A. Analyzing CRISPR Results

Instructions

DO NOT open your plates. All analysis should be performed by looking through the bottom of the closed plate.

1. Count/estimate the number of blue and white colonies on each plate. Record your data in Table 9.1.
   a. Few colonies: Count all the colonies on the plate. Use a marker to put a dot on the bottom of each plate over each colony counted (prevents double-counting).
   b. Lots of colonies: Use a ruler and a permanent marker to divide each plate into 4 regions. Count all the colonies in the quadrant. Use a marker to put a dot on the bottom of each plate over each colony counted. Multiply 4 to get an estimate of the total number of colonies on the plate.
2. Calculate the % of colonies that were edited on each plate (# white colonies on that plate / total number of colonies on the plate X 100) and add to Table 9.2.
3. Note any differences between your prediction in CRISPR Part 1 of the lab and these results in the comparison column of Table 9.2.

Table 9.2. Results of CRISPR Experiment

Plate	# Blue Colonies	# White Colonies	Total # Colonies	% White Colonies	Comparison with Prediction
A					
B					
C					
D					

Q2. Do your results support the hypothesis you made during CRISPR Part 1?

Activity 2. Perform Polymerase Chain Reaction

Do your results suggest that you were successful in performing CRISPR (turning off the *LACZ* gene)? Is there a way we can know for sure?

In this activity, we will use the process of Polymerase Chain Reaction (PCR) to check for genetic markers that indicate that our experiment was successful. During CRISPR, we added a specific sequence of DNA to the middle of the *LACZ* gene *E. coli*. If we were successful, not only should the bacteria stop turning blue, but we should be able to find that sequence in the bacterial DNA.

PCR is a process that allows us to find a specific piece of DNA in a sample and make many copies of it relatively quickly. While the overall time of this process is short, it will require a significant portion of our lab period, so we will set up our PCR reactions first and then explore the process in more depth.

Procedure A. Performing the Polymerase Chain Reaction

Follow the instructions provided by your instructor.

Activity 3. Predicting Your Results

Our PCR preparation included 3 sets of different primers:

- ❑ **Primer Set 1:** Detects unmodified *LACZ*. At one end, the primer attaches to the targeted cut site we used to perform CRISPR (the sequence encoded by the sgRNA). If CRISPR took place, the primer sequence would no longer exist, and the primer would not bind. If CRISPR did not occur, the primer would bind to that location, and the primer set would produce a 1,100-base-pair DNA fragment (amplicon).
- ❑ **Primer Set 2:** Detects modified *LACZ*. One primer in this set binds specifically to the sequence we inserted into the DNA using CRISPR. If that sequence is present, this primer set will produce a 650-base-pair DNA fragment.
- ❑ **Primer Set 3:** Control. This primer set amplifies a section of DNA that is outside of the CRISPR locus. This section is present whether or not CRISPR occurred. This allows us to ensure that PCR was performed and that it should produce a 350-base-pair amplicon in all PCR samples containing *E. coli* DNA.

Additional Key Information

The Master Mix Plus Primers (MMP) solution you added to all your PCR tubes contained all three primer sets, along with DNA polymerase and free nucleotides.

The Positive Control solution contains DNA recognized by all three primer sets and should therefore produce all three DNA fragments discussed above.

The Negative Control solution does not contain DNA.

Procedure A. Predicting Your Results

Instructions

1. Complete Table 9.2 by defining the contents and purpose of each of the PCR tubes we assembled for this experiment.
2. Use your results and the information above to complete Table 9.3.
3. In Part 3 of this project, we will perform agarose gel electrophoresis. Use Table 9. to complete the sketch of your predicted gel in Figure 9.3.

Table 9.3. Contents and Purpose of PCR Tubes

Tube	Includes DNA?	Includes Primer Mix?	Control or Experimental?	Purpose (for controls only)
S				
C				
D1				
D2				
D3				
+				
-				

	Bacterial Colony		Amplicons, bp (present/absent)		
	Colony Color	*LACZ* gene activity (+/-)	1,100	650	350
Starter Plate (IX/ARA)					
Plate C					
Plate D					

Q3. What is the purpose of the molecular weight ruler in this experiment?

Q4. Why did we perform PCR on three colonies from plate D?

Activity 4. Case File: Who Stole the Sorghum?*

Procedure A. Reading the Case Summary

Benito Fuentes (BF) and Terri Fischer (TF) are farmers who grow their crops in neighboring fields.

BF uses modern plant breeding methods to address fuel problems through agriculture. He plants a genetically modified non-sweet sorghum that produces high levels of syrup and is an excellent substrate for BioFuel production. BF closely follows all regulations for the use of genetically modified organisms and keeps his seed sealed in a storage container in his locked barn.

TF, on the other hand, grows only Traditional Food sorghum. His crop yields have been low in the past few years, and last year, he noticed that his neighbor was selling a full harvest at a great price. TF approached BF about obtaining some of his seed, explaining that he planned to breed the GMO plants with his traditional sorghum to see if he could increase the yield. BF explains that cross-breeding the two sorghum crops would be a bad idea because products from the GMO plants should not be consumed by humans, while TF's traditional crop is a human food source.

A few days later, BF's barn is broken into, and his supply of GMO seeds is taken. He reports the theft to the local police station, where his case is assigned to Martin (an expert in plant forensic science and criminology). Martin takes a statement from BF about the theft and takes notes on BF's recent discussion with TF about the seeds.

After obtaining a search warrant, the police searched TF's property and found a bag of sorghum seeds in TF's barn. TF claimed that he harvested those seeds from his own traditional crop last year. Mario determines that the only way to solve the case is with PCR and agarose gel electrophoresis

Modified from de Koff (2015) tnstate.edu and Woodrow.org (2000)

Procedure B. Modeling Polymerase Chain Reaction

Polymerase Chain Reaction (PCR) is the process by which a very small quantity of DNA is amplified (multiplied) into literally millions of copies. During PCR, only specific sections of the DNA are amplified due to the use of primers (Figure 9.1). Primers are short sequences of DNA that are complementary to a known sequence of DNA. PCR uses two primers —one on each strand —because the DNA molecule is antiparallel (Figure 9.2).

Figure 9.1. Original DNA sequence with highlighted primer sequences. The region of interest is located in between the highlighted primers.

5'-AACGAATCCCGGGATTACCAGATTACCA**AGCTT**CCTTAA-3'
3'-TTGCTTA**GGGCC**CTAATCCTCTAATGGTTCGAAGGAATT-5'

Figure 9.2. Predicted results of PCR and Agarose Gel Electrophoresis.

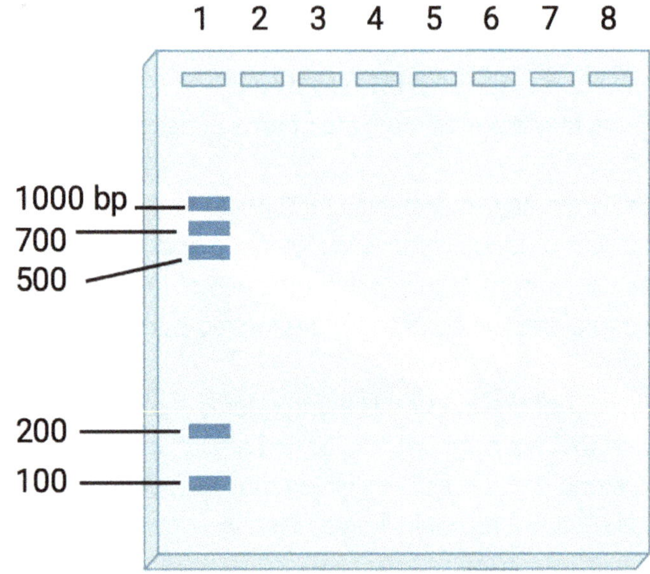

The process of PCR involves three steps:

- ❑ **Step 1. Denaturation:** Temperature is increased to 95°C to separate double-stranded DNA into two strands of single-stranded DNA.
- ❑ **Step 2. Hybridization:** The temperature is reduced to around 54°C to allow primers to attach to complementary regions on the single-stranded DNA.
- ❑ **Step 3. DNA Synthesis:** The temperature is increased to 72°C so that Taq polymerase, a DNA polymerase that works only at high temperatures, can attach to the primers and synthesize complementary DNA.

Procedure C. Performing PCR with Paper Models

In this procedure, you will perform 4 rounds of PCR using paper models.

Instructions

1. Using the paper models provided in class, cut out the double-stranded DNA molecules and distribute one to each member of your group.
2. Cut out the primers and place each one in its own pile in the middle of your workspace.
3. Cut out the blank DNA strands and pile them in the middle of your workspace.

Each group member should do the following:

4. Denaturation: Cut down the middle of the DNA molecule to separate the two DNA strands
5. Hybridization: Obtain one copy of each primer. Match the primers to their complementary regions on the DNA (remember that DNA is antiparallel!). Repeat for all DNA strands produced during denaturation.
6. Synthesis: Attach a blank DNA strip to the 3' end of each primer. Use clear tape to hold the primer, DNA sequence, and blank DNA strip together. Write the complementary base pairs onto the blank DNA starting at the 3' end of the DNA sequence.
7. Complete Table 9.5 based on the cycle you just completed.
8. Repeat until you have completed 4 cycles and Table 9.5 is complete.
9. Determine the size of the most common DNA fragment you produced in base pairs.
10. Work with your group to add your DNA to the predicted agarose gel in Figure 9.3.

Table 9.5. Who Stole the Sorghum? The Results of PCR Modeling

Cycle	# of Double Stranded DNA Fragments
Start	
After Cycle 1	
After Cycle 2	
After Cycle 3	
After Cycle 4	

Figure 9.3. Predicted gel of PCR products.

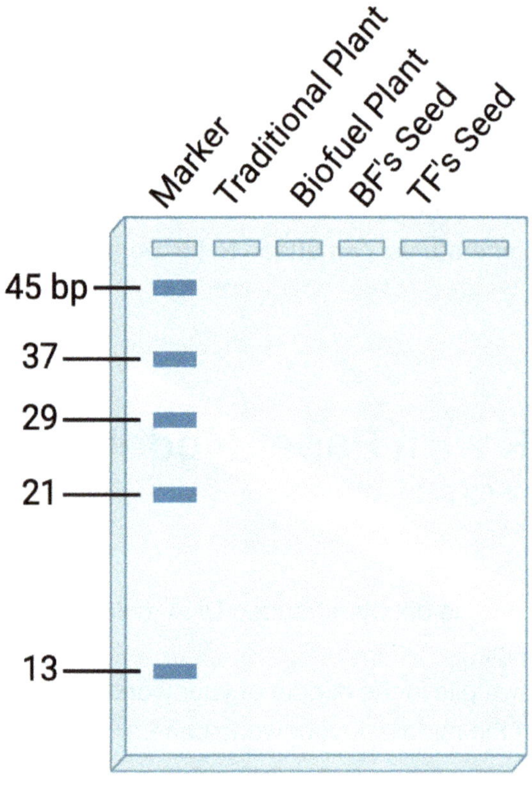

Q5. What type of "growth" did you observe during PCR based on the data you gathered in Table 9.1?

Q6. PCR usually runs for 40 cycles. Determine how many double-stranded DNA molecules you would have at the end of 40 cycles if you started with just one strand.

Q7. Why did we only use the "most common" size of DNA fragment to draw our gel?

Q8. Would the other pieces be in the sample we load into the gel?

Q9. Would they be visible?

Q10. Based on the gel, whose seeds are in the container obtained from TF's barn?

9 | Applying What You've Learned

A selection of these questions will appear on your lab quiz. Lab quizzes are open book. Type up answers to these questions in advance so that you can copy and paste the answers into the quiz.

1. Why does PCR require two primers instead of just one?
2. Crime scenes often contain very small amounts of biological material, like blood and hair, from which DNA can be extracted. Why do investigators require PCR to analyze these samples?
3. Biological samples from a crime scene are easily contaminated–they may contain DNA from two or more different individuals. How would a forensic scientist know that a sample is contaminated after performing PCR and agarose gel electrophoresis?

10 | CRISPR Out of the Blue Pt 3 Pre-Lab

Instructions

- ❏ Read the lab manual and then follow the instructions to complete the assignment.
- ❏ You may need your textbook and other resources to complete this assignment.
- ❏ Pre-labs must be completed before the lab begins.

Revisit "CRISPR Out of the Blue Part 2" to answer the following question.

1. Why did we perform PCR on our products from the CRISPR lab?

Instructions

1. Complete the following gel predicting what you should see after electrophoresis of your samples during this week's lab.
2. Double-click to open and alter the image in the drawing tool.

Figure 10.1. Predicted results of PCR and Agarose Gel Electrophoresis.

Lane	Sample
1	Molecular weight ruler (**MWR**)
2	Positive PCR control (+)
3	PCR sample **S**
4	PCR sample **C**
5	PCR sample **D1**
6	PCR sample **D2**
7	PCR sample **D3**
8	Negative PCR control (–)

10 CRISPR Out of the Blue Pt 3

In this final part of the CRISPR activity, we will use DNA electrophoresis to confirm that the DNA of the bacteria in sample D has been edited. In Part 2, we isolated DNA from several colonies and used PCR to produce multiple copies of the region of DNA around the lactase gene. The three different primer sets we used as templates for PCR produce three different sizes of DNA fragments.

In this lab, agarose gel electrophoresis is used to separate and visualize DNA fragments of different sizes in each sample.

Procedure A. Loading the Agarose Gel

Review the process of loading an agarose gel described in the Restriction Enzymes lab. Practice samples and silicon gels are provided for those wishing to practice the process before loading the actual samples.

Instructions

1. Obtain the samples produced by PCR after CRISPR Part 2 from your instructor.
2. Pulse spin your samples in the microfuge.
3. Add adaptors to the microfuge so that the small PCR tubes do not fall through the holes.
4. Add 5 µl of Loading Dye (LD) to each sample tube. Be sure to use a new tip for each sample.
5. Place the agarose gel into the electrophoresis apparatus. Remember to properly orient the wells.
6. Fill the electrophoresis apparatus with 1% TAE until the gel is fully covered in solution.
7. Load your samples into the wells as shown in Table 10.1.

Lane	Sample	Amount (µl)
1	Molecular Weight Marker (MW)	20
2	Positive Control (+)	15
3	Sample S	15
4	Sample C	15
5	Sample D1	15
6	Sample D2	15
7	Sample D3	15
8	Negative Control (–)	15

8. Run the gel at approximately 200 mV until the blue dye is 3 to 5 cm from the wells.
9. Use the UV light box to view your gel. Photograph or draw a sketch of the gel for later analysis.

Q1. Were the results in each lane of your gel what you expected? Why or why not?

Q2. As a whole, what do these results tell you about CRISPR and the experiment we performed in Part 1 of this lab?

10 | Applying What You've Learned

A selection of these questions will appear on your lab quiz. Lab quizzes are open book. Type up answers to these questions in advance so that you can copy and paste the answers into the quiz.

1. We used PCR and agarose gel electrophoresis to confirm that the colonies were white in the D samples because the DNA was edited. What would we have seen in the gel if the white phenotype was not a result of CRISPR?

Appendix A. Reference to Basic Microscope Techniques

Compound Microscope Anatomy

Using a microscope effectively requires knowing a bit about how it works. An excellent resource to learn about microscope function is to explore the virtual microscope using the QR Code 1A.

QR Code 1A. Exploring a Virtual Microscope

https://www.ncbionetwork.org/iet/microscope/

Figure A1 contains a labeled diagram of the microscope used in this class, and the remainder of this section provides a step-by-step procedure for using the microscope.

Figure 1A. Labeled diagram of the Olympus Student Microscope.
Numbers correspond to those in Figure 1A.

1. Base
2. Arm
3. On/Off switch
4. Rheostat
5. Objectives
6. Stage
7. Mechanical stage feet
8. Iris diaphragm
9. Coarse focus knob
10. Mechanical stage adjustment knobs
11. Fine adjustment
12. Ocular
13. Condenser
14. Condenser adjustment knob
15. Lab shutte

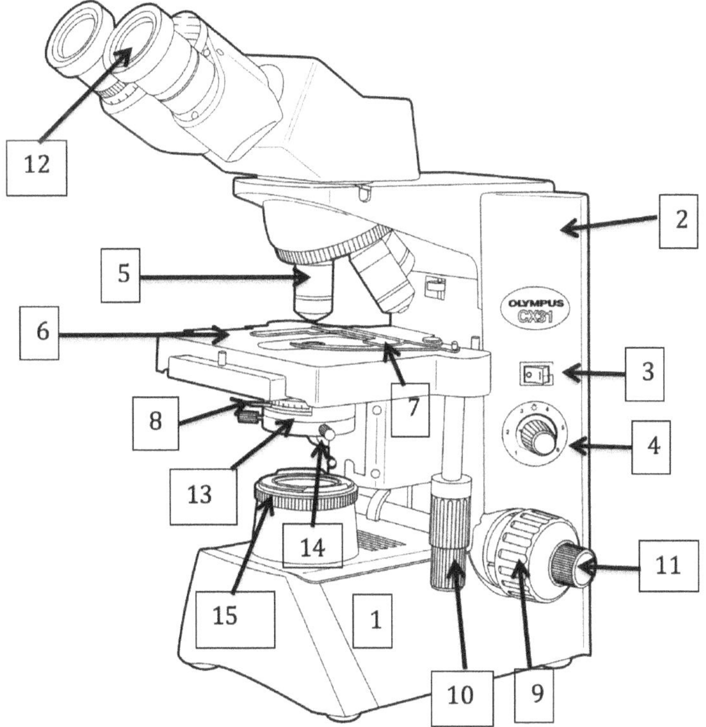

Using a Compound Microscope

Instructions

1. Remove the microscope from the cabinet with two hands, one under the base (1) and one in the handle on the arm (2). Place the microscope on the counter, remove the dust cover and store it in the cabinet.
2. Plug the microscope in and turn on the light (3). Set the rheostat (4) that controls the light brightness, turn it up to about "5" .
3. Ensure that the scanning (smallest lens and lowest power) objective (5) is rotated into place over the hole in the stage. If not, rotate the nosepiece until you feel the objective click into place.
4. Place your slide on the stage (6) between the metal "feet" (7). The slide should go between (not under the feet) with the specimen on top and in the center of the hole in the stage.
5. Reduce the brightness by adjusting the diaphragm (8; the black plastic lever). Move the lever all the way to the right, so that it appears darkest when looking through the microscope, then slowly move the lever to the left until it just stops getting brighter. This should provide you with the greatest balance of contrast and resolution while viewing slides.
6. Raise the stage using the coarse focus knob (9) until it is as close to the slide as possible. Ensure that the lens does not touch the slide.
7. Focus by rotating the coarse focus knob while looking through the eyepiece. You will lower the stage until you see a sharp image.
8. When the image is in focus, use the mechanical stage adjustment knobs (10) to move the slide so that the part you want to see is exactly in the middle of the lit area (field of view).
9. Use the fine focus knob (11) to sharpen the image.
10. If you want to see greater detail, rotate the nosepiece so that the low-power objective (10X) clicks into position. Adjust the focus with the fine focus knob. DO NOT LOWER THE STAGE BEFORE YOU SWITCH OBJECTIVES. If the image is focused on one objective, the next objective will not hit the slide.
11. Go through the same process of focusing and centering before you go to high power (40X).

Putting Away Your Microscope

Don't forget to clean up. Microscopes require special care when being stored. Check the following:

- ❑ Rotate the nosepiece until the shortest (scanning) objective is in place and lower the stage to its lowest point.
- ❑ Remove the slide from the stage, clean and dry it, and put it back where you got it.
- ❑ Clean off any water on the stage with a paper towel or absorbent paper.
- ❑ Coil the cord on the back of the microscope, and put the dust cover on.
- ❑ Return the microscope to the cabinet with its oculars facing away from you.

Preparing a Wet Mount

A wet mount is a slide of freshly prepared material made quickly in a laboratory setting. In general, wet mounts include living material that is not stained or treated with preservatives.

Instructions

1. Obtain a clean slide and cover slip.
2. Place a drop of liquid (water or saline solution depending on the type of organism you are examining) in the center of the slide.
3. Place the specimen in the drop of liquid. The specimen should be very small and as thin as possible.
4. Hold the cover slip between your thumb and forefinger. Place one edge of the cover slip on the slide near the left side of the liquid containing your specimen. The cover slip should be at a 45° angle to the slide.
5. Slowly lower your cover slip onto the slide.

Staining Wet Mounts

Cells can be difficult to see as they often lack color and contrast. Adding a stain can help you differentiate between cell parts and can sometimes identify specific structures within the cell. To prevent overstaining, stains are added to wet mounts by capillary action rather than directly to the specimen.

Instructions

1. Place a drop of the stain on the slide, touching the edge of the cover slip.
2. Twist the corner of a tissue into a point. Place the point at the edge of the other side of the cover slip (opposite the drop of stain).
3. Capillary action will draw the stain under the cover slip.
4. E. Using the field of view diameter to estimate specimen size.

When you look into a microscope, you see a circle of light called the field of view (Figure 2A). As you go up in magnification, the area of the specimen you can see without moving the slide gets smaller.

We say that the field of view decreases as magnification increases, and you can visualize this change in size by measuring the diameter of the field of view at each magnification with a ruler (just put a clear ruler on the stage).

Since the objectives of your microscope don't change, the diameter of the field of view at each magnification should remain the same as long as you are using the same brand of microscope.

Figure 2A. Defining the field of view.

Table 1A includes the field of view measurements for the microscopes used in this class. Micrometers (10-6 m) are the unit of measurement used by microscopists.

Table 1. Field of View at Different Magnifications on the Olympus Microscope

Objective	Diameter of Field of View
4X	2500 µm
10X	1000 µm
40X	500 µm

Now that you have calibrated your microscope, you can use those measurements to estimate the size of the specimens you observe.

1. Bring a specimen into focus with the lowest power objective and center it within the field of view.
2. Determine the best magnification to use when measuring your specimen.
3. Choose an objective where the field of view is as filled by the object you are measuring, but the object should not be bigger than your field of view.
4. Choose a magnification that provides the level of detail you require for observations. Too much magnification often makes the image hard to interpret.
5. Choose one dimension (height or width) of the object you are measuring.
6. Determine the fraction of that field of view that the object occupies (see Figure 3A).
7. Multiply that fraction by the known diameter of the field of view to obtain the estimated size of the object.

Figure 3A. Creating a scale bar for your notebook.

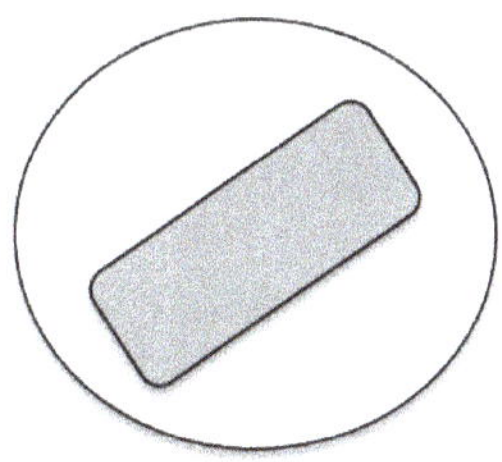

Microscope view
Field of view = 100µm

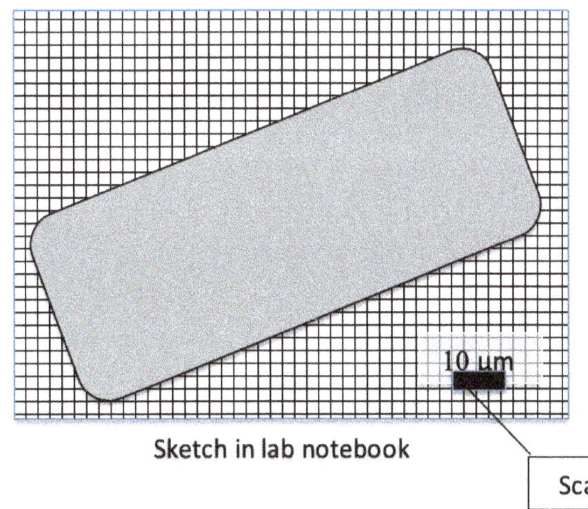

Sketch in lab notebook

Scale bar

Based on the size of the field of view (left), the cell is estimated to be 80 μm long. In the sketch (right), the cell is about 25 squares long. 80 μm/25 squares = 3.2 μm/square.

You could draw a sketch bar that is 1 grid square long and write 3 μm over it. To make the scale bar easier to see here, it was drawn 3 grid squares long, which is approximately equivalent to 10 μm.

Creating Scaled Sketches

Researchers keep detailed records of what they see and do in the laboratory both as reference and to support any publications they produce. Accuracy is important when creating these records. While photographs can provide records of what is seen in a microscope, sketches often provide more detailed observations.

When creating sketches, it is important that the size of the organism is accurately represented. Since we can estimate the size of an object with the microscope, we can also create scaled sketches.

Instructions

1. Create a sketch of the organism you are observing on graph paper.
2. The sketch should fill at least ⅙ of a page of standard-sized paper.
3. Add as much detail to the sketch as possible.
4. Label any structures visible in the sketch.
5. Use a section ⅙ of the paper to record any observations you made about the specimen.
6. After you have finished sketching, return to the microscope to estimate the length or width of the object you sketched. Jot that number down in the observations section of your sketch.
7. Create a scale bar for your drawing (see Figure 3A for an example).
8. Count the number of grid squares spanned by the measured dimension in your sketch
9. Divide your estimate of the cell size based on the field of view by the number of grid squares spanned by your sketch to get μm/grid square.
10. Fill in 1-2 grid squares to create a scale bar.
11. Print the conversion factor (calculated in B) above the scale bar.
12. This is an estimate! Round to a whole number.
13. The value of a scale bar should be 1 μm or a multiple of 5 μm.

Appendix B. Writing a Scientific Paper

Communication is a key part of being a scientist. Scientists rely on feedback to decide what questions to answer, design hypotheses, design better experiments, interpret their data, and draw conclusions. Scientific communication occurs through a variety of processes, ranging from general discussions at lab meetings to the publication of scientific papers. In this exercise, we focus on the key steps involved in writing a scientific paper.

A scientific paper typically includes five main parts:

- ❑ Introduction
- ❑ Methods
- ❑ Results
- ❑ Discussion
- ❑ Conclusion

Equally as important are the **title** and **references** section. Most scientific papers will also include an abstract (we will not work on those in this class). Each part of the paper has specific criteria associated with it, and those criteria may vary depending on the journal you choose to publish your paper in. As a result, *one of the key aspects of scientific publication is following directions*. The criteria for each part of a scientific paper are reviewed in the sections below.

Working with Published Material

Scientific writing requires the use of published material to support your understanding and interpretation of data. Most of the sources you will use will be primary sources. Primary sources focus on a specific data set obtained from a single source. These publications report observations and attempt to interpret them. In our class, primary sources are always peer-reviewed, meaning they have been examined by experts in the field and are considered to represent "good science."

Secondary sources are those that provide overviews of many primary sources, summarize trends, and present well-established hypotheses. Secondary sources are low in detail, but their reference sections are a goldmine! Use secondary sources to get an overview of the topic and their reference sections to find more detailed information about the topic you are interested in. Note that secondary sources often appear in peer-reviewed journals where they are called "reviews"–even though these articles are published in a journal, they are not considered primary sources.

In this class, you will use published material in two ways: to substantiate facts used in your writing and to support your own arguments and interpretations. The way to incorporate a source into your writing depends on how you intend to use it.

Using Published Material to Substantiate Facts

When to Use

- ❏ Background section of the introduction
- ❏ When stating a fact or explaining a concept that is not covered in your textbook
- ❏ Whenever a specific number or date is used in the text
- ❏ When discussing species-specific information
- ❏ When new information is brought into the discussion

How to Incorporate

- ❏ In-text citation (see references section of this document)

Using Published Material to Support Arguments and Interpretations

When to Use

- ❏ In the introduction, when explaining how you came up with your hypothesis
- ❏ In the discussion, when explaining what you think your data means

How to Incorporate

- ❏ Write a summary of the study performed. What was the experiment, what were the results, and how are those results relevant to your work?
- ❏ Include an in-text citation at the end of your description.
- ❏ If the description is long, you may need to include a citation at the beginning and end of your summary.

Title

The title of a paper seems like an insignificant thing. Who cares about the title as long as the rest of the paper is good, right? Think again! Titles can be one of the most important parts of a scientific paper. Generally, scientists publish their papers in scientific journals where papers are listed by title in the table of contents. Your goal as a scientist should be to create an informative but interesting title that will encourage the readers of the journal to turn to the first page of your article.

Writing a Title

In terms of content, the title of a scientific paper should summarize in one sentence or less the overall theme of the paper, including the independent and dependent variables. Indicate the question being answered and hint at the conclusion. Beware of titles that seem too short or too long!

Fava Bean Growth

What aspect of fava bean growth interests you? Why is fava bean growth interesting? Why would I want to learn more about it?

Long Titles are Generally Repetitive

Organic Fertilizers including Manure and Chemical Fertilizers like Rapid Grow Cause Fava Beans to Grow More Rapidly than Untreated Fava Beans but Chemical Fertilizers Do Not Cause Fava Beans to Grow Faster than Organic Fertilizers

Did you make it through that one? It's challenging to capture and keep a reader's attention with such a long title. Not only does this title provide too much information, but it is also very repetitive.

Try This Title

Organic and Chemical Fertilizers are Equally Efficient at Increasing the Growth of Fava Bean (Vicia faba) Seedlings

Long enough to tell you what's going on and imply the question the paper is focusing on, but short enough that a reader can get through it. That's your goal.

General Suggestions for the Title

- ❏ Only the first letter of the title and proper nouns should be capitalized.
- ❏ Usually, scientific names (not common names) are listed in the title. Remember, a scientific name is always underlined or italicized. Genus is always capitalized, and species is not.
- ❏ The title should be a slightly larger font size than the rest of the text and can be highlighted by using bold type.

Introduction

This section of the paper provides the reader with an Introduction to the topic being discussed and focuses the reader on specific aspects of this topic. The introduction ends with a statement of your hypothesis and a prediction of what your results will be.

Introductions begin with general information and become very specific toward the end. The text should begin with basic background information and work slowly toward the specifics of the research. Assume the reader is completely unfamiliar with the topic, and you need to provide them with all the information they need to understand the experiments. Any information provided should come from published literature that you cite in the text and list in your references section.

> Originally, farmers treated their fields with natural amendments (organic fertilizers) such as compost, manure, and in some cases, dead fish (Smillee, 1997). In the last 50 years, synthetic fertilizers have become a common replacement for these natural fertilizers (Smith and Jones, 1995). Fertilizer companies claim that synthetic fertilizers deliver nutrients to plants more efficiently than organic fertilizers, thereby helping plants grow larger and increasing crop yields (Gonzco et al., 1998).

The next section of the introduction should prepare the reader for what you plan to do. What is your question? How will you study that question, and what do you think will happen? At the end of this section, include your hypothesis and predictions as to how your experiments will turn out.

> Sam and Green (2001) found that 63% of the nitrogen in a synthetic fertilizer is released within the first hour that the fertilizer is exposed to water. Organic fertilizers release nutrients into the soil at a much lower rate. Commercially available compost was shown to retain 99% of its nitrogen content a week after application and watering (Feddle, 1999). We propose that sustained release of nutrients from organic fertilizers provides longer-term benefits regarding seedling growth.

Notice that references are provided in this paragraph to support the reasoning behind the hypothesis. Your introduction should include two sets of references: those that provide essential background information and those that help explain how you derived your hypothesis.

General Suggestions for the Introduction

- ❑ Always write in the past tense.
- ❑ Avoid using the first person, except when discussing your hypothesis and making predictions.
- ❑ Be generous in your explanations. Make sure the reader has the background to understand everything you will be doing in your paper.
- ❑ Move from general to specific.
- ❑ Make smooth transitions as you move to more specific material. What does the second paragraph have to do with the first? Tie it all together!

Methods and Materials

The Methods section provides the reader with the procedure you performed. This may not seem like a critical section, but it can make or break a paper. When a colleague reads a paper, they carefully analyze the methods section to ensure that techniques and machinery were used correctly and that the proper controls were performed. Some scientists may even try to replicate your experiments.

The Methods section is more than simply rewriting the instructions; it also includes additional details. You should provide enough information that your experiment can be reconstructed, but not the gory details. You want to provide the reader with enough information to repeat the experiment, but you don't want to waste precious words. Remember that your laboratory manual is written in a simplified format (like a cookbook) so beginning biology students can easily follow the instructions. More experienced scientists will be able to infer a certain amount of information from what you write.

Sample Methods and Materials Section

To study the effects of synthetic and organic fertilizers on seedling growth, fava beans (*Vicia faba*) were planted in sterilized sand with or without amendments. Twenty-five grams of synthetic fertilizer (Rapid Grow granules) was mixed with 500 g of sand for the "synthetic fertilizer" treatment. Twenty-five grams of composted manure was mixed with 500 g of sand for the "synthetic fertilizer" treatment.

Seeds were soaked in water for 12 hours prior to planting and planted at a depth of 2.5 cm and at a density of 25 seeds per flat. Germination and growth occurred in a greenhouse maintained at 25°C with a 14-hour light cycle followed by a 10-hour dark cycle. Seeds/seedlings were watered daily. Seedling height was measured daily after the emergence of shoots was observed (approximately 10 days after planting).

General Suggestions for Methods & Materials Section

- ❑ Never use the first person ("I", "we", "our", etc.).
- ❑ Always use the past tense.
- ❑ Do not include the date and time you performed the experiment.
- ❑ Do not include descriptions of containers (like beakers) or equipment (like microscopes) unless they are unique and their design is essential to the experiment.
- ❑ Do not explain how to do standard laboratory protocols (making a solution, determining the mass of a sample).
- ❑ Do not include labeling or labeling methods unless they are key to the experiment.
- ❑ Do not state that you recorded the data.
- ❑ Be sure your text includes variations in treatments (including the control). What chemical concentrations did you use? How long was the treatment? What was the temperature?
- ❑ Clearly define your dependent variables.

Results

The Results section is where you present your data to the reader in a completely objective manner. Separating the data and your interpretation of the data allows readers to draw their own conclusions, but you don't want your readers to draw just any conclusion. You want them to conclude the same thing that you concluded. To accomplish this, your writing must act as a guide for the reader, highlighting key pieces of data and effectively leading the reader to the conclusion you want them to draw. The Results section of a paper gives the reader **only the facts** of the experimental results. The facts are presented in three ways: figures, tables, and text.

Tables & Figures

A scientific paper presents data in two formats: tables and figures. A table is a chart based on rows and columns where data, numeric or otherwise, is organized by category. A figure is any type of visual data display that does not involve a table. Figures include graphs, photographs, diagrams, and drawings. Figures and tables are numbered separately in a formal scientific paper.

Tables vs. Graphs

Some of the types of figures are obvious, but you may find the use of tables and graphs confusing. We often collect data in the form of a table because it is easy to keep the data points organized; however, this is not usually the best way to visualize how the dependent variable is affected by a change in the independent variable.

Use a Table When:

- ❏ Neither your dependent nor your independent variable is quantitative
- ❏ You are displaying single data points for each treatment
- ❏ Your overall number of data points is small (5 or fewer)
- ❏ When the data points are not continuous

Use a Graph When:

- ❏ At least one of your variables is numeric
- ❏ Your data points are continuous
- ❏ You are displaying more than one independent variable
- ❏ You want to show a trend or rate of change
- ❏ You have many data points (more than 5)

Bar Graphs vs. Line Graphs

If you have decided to display your data in a graph, you have one more decision to make. What type of graph should you create? Many types of graphs exist, but most often, you will need to choose between a bar graph and a line graph.

Use a Line Graph When:

- ❏ Both variables, independent and dependent, are numeric
- ❏ The data is continuous
- ❏ The differences between data points occur on a small scale (you can adjust the scale on the axis to make the differences more obvious)
- ❏ More than one independent variable is being plotted on the same graph

Use a Bar Graph When:

- ❏ One of your variables is not a numeric value
- ❏ The data is non-continuous
- ❏ The data is continuous, and the change between data points is significant
- ❏ Categories need to be displayed in relation to one another

> Notice that there is some overlap between these graph types. The key is to determine which format best displays the trends you see in the data.

Graph Format Requirements

- ❏ The **independent variable** always goes on the X-axis.
- ❏ The **dependent variable** always goes on the Y-axis.
- ❏ Choose a scale that is appropriate for each axis (your data points should fill your graph rather than being squashed together at one end).
- ❏ The X- and Y-axes should be clearly labeled, and units of measurement should be indicated.
- ❏ If a graph has more than one line, the different lines should be clearly distinguishable by color and/or line texture, and a key should be provided.
- ❏ Numbers should not be attached to data points.

All figures and tables should include a **descriptive legend** that includes a brief description (1–2 sentence) of the table or figure. The description should be focused on what the reader needs to know to understand the figure. That usually means it contains information about how the data was prepared for display. A guide to data points and the results of statistical analysis is often mentioned in the figure legend. The key to allowing your text and figure to both integrate AND stand alone is to provide a strong figure legend. **Figures and Tables should be labeled separately**, and the number should reflect the order in which the figure is used in the results section.

Results Section Text

The text portion of the Results section should provide your observations in text format. Make sure your descriptions are thorough. Try to "paint a picture" of your data with your words. If someone has a copy of your paper but loses the figures, they should be able to form a mental image of the figures based on the description in the text.

- ❑ Refer to all your figures in the text of this section. Every figure you use in your paper must be mentioned somewhere in your text!
- ❑ **Do not provide any opinions on your data in this section.**
- ❑ **NEVER CHANGE YOUR DATA.** Even if you think it's wrong!
- ❑ Remind your reader what the experiment is about. Why did you do this particular experiment?
- ❑ Describe what you saw during the experiment. Did the sample change color? Temperature? Consistency?
- ❑ Point out any trends in the data (the temperature increased throughout the experiment).
- ❑ Point out any surprising data points. You can later explain these in the discussion section.
- ❑ Statistical analysis of your data should be included in the results section.

Sample Results Section

To examine the effect of organic and synthetic fertilizers on seedling growth, fava beans (*Vicia faba*) were planted in unfertilized sand, sand amended with synthetic fertilizer, and sand amended with organic fertilizer. After germination, seedling height was measured daily.

Figure A.1. Sample Results

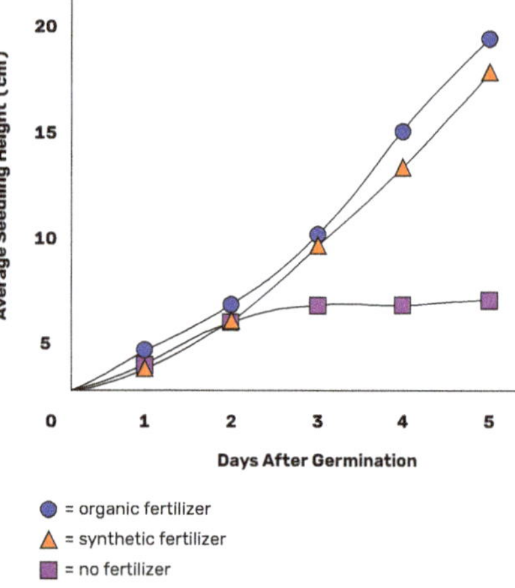

The seedlings that were given different fertilizers grew faster than those that were unfertilized, but the rate of seedling growth differed little between the two types of fertilizer (figure A.1). The average seedling height 5 days after germination was 19 cm in the plants fertilized with organic fertilizer and 17 cm in the plants fertilized with organic fertilizer. Unfertilized plants grew to approximately 5 cm. The rate of growth is constant for both organic and synthetic fertilizers, as indicated by the similar slopes of the lines for these treatments in figure A.2. Unfertilized plants initially grew at the same rate as fertilized plants but stopped growing after Day 3.

Discussion

The art of science is examining the data and interpreting it. Sometimes that interpretation will be that we believe the experiment was poorly executed; other times it will be an exciting new discovery. In the Discussion section, it is your job to interpret your data.

- ❑ Interpretation involves explaining why you think you got the data you did, not just drawing a conclusion.
- ❑ Use what you know about how life works and how your system works to make sense of the results.
- ❑ Lead the reader to your conclusion rather than allowing them to draw their own conclusion.

Begin your discussion by restating your hypothesis and explaining whether your data supports or refutes the hypothesis.

> We proposed that the application of organic fertilizers would increase the growth rate of fava beans more than synthetic fertilizers. Although both types of fertilizer increase the rate of seedling growth compared to unfertilized seedlings, fava bean seedlings treated with synthetic or organic fertilizers grew at approximately the same rate and were approximately the same size after 5 days (Figure 2). This does not agree with previous results that show organic fertilizers produce larger crop yields in tomatoes (Smeagol et al., 1998) and strawberries (Hallo and Schmidt, 1999). Unfertilized seedlings stopped growing after 3 days.

The meat of the discussion really has two parts. First, explain why you think you got the results that you got. Second, use previously published scientific evidence to support that explanation.

- ❑ You do not need data that exactly matches yours.
- ❑ You do need to demonstrate that your interpretation is scientifically valid.

> During germination, storage proteins, lipids, and carbohydrates stored in the cotyledons of the fava bean are used to power and support the growth of the seedling. Once photosynthesis begins, seedlings absorb water and inorganic nutrients from the soil. Nea and Popper (2012) used radioactive nitrogen to track the flow of nutrients into sunflower seedlings. They found that 3 days after photosynthesis began. At the same time, radioactive nitrogen was still present in seedling cotyledons; it ceased moving from the cotyledon into the seedling and began to flow into the seedling from the soil. This explains why seedlings grown in sand without fertilizer stopped growing after 3 days. As the seedlings in this experimental group depleted their stored nutrients, they were unable to absorb additional nutrients from the soil (since it lacked all forms of fertilizer). They could no longer obtain the building blocks necessary for synthesizing new plant tissues.

The similarity in the results of the two fertilizer treatments may be a result of the concentration of fertilizer used. Synthetic fertilizers contain more concentrated nutrients than organic fertilizers (Porter, 2011); however, in this experiment, we used the same amount (by weight) of both types of fertilizers. Determining the actual concentration of key nutrients, such as nitrates and phosphates, in synthetic and organic fertilizers would allow us to apply them at the same concentration to the soil and ensure that differences between treatments were due to the type of fertilizer, rather than the concentration of the fertilizer.

Then, address any limitations in the experiment(s). Limitations are not necessarily errors or flaws. You can only investigate one small piece of the puzzle in any one experiment. What new experiment would expand your ability to understand this process and/or make your results more applicable?

These results were limited by the fact that only one type of organic fertilizer was tested. Most organic farmers use a combination of three to five different fertilizers to produce maximum crop yield (Cenote, 1999). Different organic fertilizers provide different types of nutrients. Bone meal, for example, is high in phosphorus and has some nitrogen (Cenote, 1999). Future experiments should be designed to examine seedling growth in Fava beans treated with combinations of manure, bone meal, bat guano, and alfalfa meal, ensuring that seedlings receive the optimal combination of nutrients.

Finally, end the discussion section with a concluding paragraph.

Our results suggest that organic and chemical fertilizers are equally effective in promoting plant growth. However, further research is necessary to determine the optimal concentration of fertilizer required and whether other organic fertilizers or combinations of organic fertilizers will have a similar effect.

> Notice that you can use the first person in the discussion section. It is best, however, to limit its usage. The data should stand on its own—you're just interpreting it!

General Suggestions for the Discussion Section

- ❑ Always use the past tense.
- ❑ Refer (again) to all your figures as you explain them.
- ❑ Use peer-reviewed references to support what you say.
- ❑ Assume your reader knows nothing. At this level, it is essential for you to explain exactly why you got the results you did (or why your results did not match your hypothesis). For example, if heat increases the rate of a reaction, explain why (at the molecular level).
- ❑ Ensure that you interpret all your data accurately. For instance, if there's a flat portion of a graph, tell your reader why it is there or at least provide a hypothesis. Explain whether you expected these results and why or why not.

References

Any information that is not considered **standard knowledge** must be cited and referenced in a scientific paper. As a general rule, consider anything found in your biology textbook as standard knowledge in the field of science. Wherever you go into more depth, you need to provide a reference.

Scientists do not quote directly from papers but summarize the facts or conclusions of a paper in one or a few sentences that are referenced at the end of a sentence. You must practice summarizing previously published material in your own words. **Points will be taken away from your grade for using quotations rather than summarizing your references.**

In-text Citations

When you use a reference, you must demonstrate that use in two ways. First, provide an abbreviated **in-text citation** in the body of the paper. Second, follow up that citation with a complete and detailed **reference** in the reference section of the paper. *The citation should match the beginning of the reference* so that the reader can easily move back and forth between the paper and the reference list when necessary.

Format: (Author Last Name, Year)

- ❑ Scientific citations include the author and the year of publication, but not the page number.
- ❑ Citations are placed in parentheses at the end of a sentence before the period.
- ❑ If you mention an author in the text, follow that author with the year of publication in parentheses ("According to a study by Smith (2006), …").
- ❑ If there are more than two authors for a paper, abbreviate the list by using the first author's last name followed by "et al." (Jones et al., 2002).
- ❑ If one citation covers an entire paragraph, either place the citation at the end of each sentence or mention the study at the beginning of the paragraph and provide the citation there.

Reference List (Works Cited)

Your reference list links your in-text citations to a complete reference, directing the reader to more information and providing verification for your discussion. Each science journal has a specific set of formatting requirements. As an example, this class has its own specific requirements. The entire, final list of references should be **alphabetized** (by author's last name, which should come first) so that your reader can easily match up each in-text citation with the full citation in the list. **Templates and sample references are provided below.**

Journal articles (including those accessed digitally)

Author(s)(year) Title of article, *Journal name* Volume: pages.

Freas KE, PR Kemp (1983) Some relationships between environmental reliability and seed dormancy in desert annual plants, *Journal of Ecology* 71: 211–217.

Books

Author(s) (year) *Book title*, Publisher, City, State, and country where the publisher is located.

Fenner, M (1985) *Seed ecology*, Chapman and Hall, New York, NY, USA.

Edited Books

Author(s) (year) Chapter title. In *Book Title* (names of editor(s), eds.), Publisher, City, State, and Country, pages.

Kemp PR (1989) Seed banks and vegetation processes in deserts. In *Ecology of Soil and Seed Banks* (MA Leck, VT Parker, RL Simpson, eds.), Academic Press, New York, NY, USA. 257–282.

Websites (DO NOT Include in Final Paper)

Author(s)[1] (year[2]) Title of Webpage. Retrieved on <date you visited that website> from <complete website URL>.

Backyard Gardener (no date) Seed Germination Database. Retrieved on February 8, 2011, from http://www.backyardgardener.com/tm.html.

---

[1] Find the author at the very bottom of the page. If no author is listed, use the sponsor of the website (check the URL)
[2] Some webpages lack a date of publication. If this is the case, put "no date" in the parentheses.

General Suggestions for the References Section

- ❑ Make sure that the in-text citation matches the start of the associated reference in the reference list. This is how the reader connects the citation to the actual source.
- ❑ Italicize all scientific names used in the references section.
- ❑ Book and Journal titles are always italicized or underlined.
- ❑ Chapter titles and article titles are in normal font, and only the first letter of the first word is capitalized.

Peer Review & Publication

Scientists do not work alone; they work as part of a team. Although there is some competition in science, most scientists prefer to work with their colleagues by sharing their ideas and data and receiving feedback. The publication of scientific papers in a respected journal is a crucial step in the scientific process, as it makes your work accessible to the public. Once a paper is published, a larger number of scientists are available to provide additional feedback and ideas to further your research.

Peer review is a process that ensures the validity of your work. After submitting a scientific paper to a journal, the journal will identify experts in the field related to the paper and send it to them for critique. When you read a paper that has been published after peer review, you can be confident that the logic, procedure, and ideas presented have been thoroughly vetted by a group of experts who are familiar with that topic. That being said, reviewers do make mistakes! Occasionally, a paper that has passed through the peer-review process is later retracted.

Throughout the Principles of Biology series, you will be asked to peer-review the work of your teammates and other members of your class. Being able to provide constructive criticism in these reviews (and being able to handle the criticism you receive in return) is all part of the scientific process.

Tips on Providing Constructive Criticism

- ❑ **Remember that your goal is to help the writer produce a better product**, rather than making them feel that they are unprepared or unskilled.
- ❑ **Provide both praise and critique in your comments.** Point out to the writer what you like about their work before you start addressing your concerns.
- ❑ **Be specific in your praise.** Telling the writer that you like their work overall will not mean as much as when you say you liked a specific part of the introduction. Being specific shows that you really read the paper.
- ❑ **Be specific in your feedback.** Instead of saying that the introduction doesn't make sense, point to a particular paragraph or section that confuses you and explain why.

- **Ask questions as well as making comments.** Whenever possible, use comments like "Did you mean…" or "Would …be a better word to use here?" Questions are a less confrontational way to get your point across and provide the opportunity for the writer to correct things on their own.
- **Remember that it's okay to admit you don't know how to fix something.** Point out your concern, but acknowledge to the writer when you don't have a quick fix.

Tips on Receiving Constructive Criticism

- **Read all of the feedback provided before reacting**, then sit back and let it sink in for a minute.
- **Don't take the feedback personally.** Remember that your reviewer is trying to help you get a better grade, not hurt you.
- **Remember that you don't have to do everything** suggested by your reviewer. We use more than one reviewer so we can get different opinions. If all your reviewers say the same thing, then you should take the comment seriously, but if only one person commented on a specific item, ask yourself whether the comment is key to your success or if it is a point of personal preference on the part of the reviewer.

Acknowledgments

Unless otherwise stated, figures are created and copyrighted by the authors or Chemeketa Press. © 2025.

Lab 1

Figure 1.2. "Human Erythrocytes OsmoticPressure PhaseContrast Plain" by author Zephris. Wikimedia Commons Creative License 3.0.

Lab 2

Figure 2.1. "Amylase reaction," by BQmUB2012134. Wikimedia Commons Creative License 1.0.

Lab 5

Figure 5.2. "Apical Meristem in Onion Root Tip (34886438470)" by Berkshire Community College Bioscience Image Library. Wikimedia Commons Creative License 1.0.

Lab 7

Figure 7.1. "Dreistachlige Stichling Gasterosteus aculeatus" by Holger Krisp. Wikimedia Commons Creative License 4.0.

Appendix A

Figure A.4 "Parts of a Microscope (english)" by Thebiologyprimer. Wikimedia Commons Creative License 1.0.